Lecture Notes in Computer Science 6360

Commenced Publication in 1973
Founding and Former Series Editors:
Gerhard Goos, Juris Hartmanis, and Jan van Leeuwen

Maristella Agosti   Nicola Ferro
Carol Peters   Maarten de Rijke
Alan Smeaton (Eds.)

Multilingual and Multimodal Information Access Evaluation

International Conference
of the Cross-Language Evaluation Forum, CLEF 2010
Padua, Italy, September 20-23, 2010
Proceedings

 Springer

Volume Editors

Maristella Agosti
University of Padua, Italy
E-mail: agosti@dei.unipd.it

Nicola Ferro
University of Padua, Italy
E-mail: ferro@dei.unipd.it

Carol Peters
ISTI – CNR, Pisa, Italy
E-mail: carol.peters@isti.cnr.it

Maarten de Rijke
University of Amsterdam, The Netherlands
E-mail: derijke@uva.nl

Alan Smeaton
Dublin City University, Ireland
E-mail: alan.smeaton@dcu.ie

Managing Editor

Pamela Forner
CELCT, Trento, Italy
E-mail: forner@celct.it

Library of Congress Control Number: 2010934520

CR Subject Classification (1998): I.2.7, H.2.8, I.7, H.3-5, H.5.2, I.1.3

LNCS Sublibrary: SL 3 – Information Systems and Application, incl. Internet/Web
and HCI

ISSN	0302-9743
ISBN-10	3-642-15997-4 Springer Berlin Heidelberg New York
ISBN-13	978-3-642-15997-8 Springer Berlin Heidelberg New York

springer.com

© Springer-Verlag Berlin Heidelberg 2010
Printed in Germany

Typesetting: Camera-ready by author, data conversion by Scientific Publishing Services, Chennai, India
Printed on acid-free paper   06/3180

Preface

In its first ten years of activities (2000-2009), the *Cross-Language Evaluation Forum (CLEF)* played a leading role in stimulating investigation and research in a wide range of key areas in the information retrieval domain, such as cross-language question answering, image and geographic information retrieval, interactive retrieval, and many more. It also promoted the study and implementation of appropriate evaluation methodologies for these diverse types of tasks and media. As a result, CLEF has been extremely successful in building a wide, strong, and multidisciplinary research community, which covers and spans the different areas of expertise needed to deal with the spread of CLEF tracks and tasks. This constantly growing and almost completely voluntary community has dedicated an incredible amount of effort to making CLEF happen and is at the core of the CLEF achievements.

CLEF 2010 represented a radical innovation of the "classic CLEF" format and an experiment aimed at understanding how "next generation" evaluation campaigns might be structured. We had to face the problem of how to innovate CLEF while still preserving its traditional core business, namely the benchmarking activities carried out in the various tracks and tasks.

The consensus, after lively and community-wide discussions, was to make CLEF an independent four-day event, no longer organized in conjunction with the *European Conference on Research and Advanced Technology for Digital Libraries (ECDL)* where CLEF has been running as a two-and-a-half-day workshop. CLEF 2010 thus consisted of two main parts: a peer-reviewed conference – the first two days – and a series of laboratories and workshops – the second two days.

The conference aimed at advancing research into the evaluation of complex multimodal and multilingual information systems in order to support individuals, organizations, and communities who design, develop, employ, and improve such systems. Scientific papers were solicited in order to explore needs and practices for information access, to study new evaluation metrics and methodologies, and to discuss and propose new directions for future activities in the multilingual and multimodal *Information Retrieval (IR)* system evaluation in context. In addition, we encouraged the submission of papers that analyzed some of the achievements made during 10 years of CLEF through in-depth experiments using existing CLEF collections in imagintive and interesting ways. A large program committee, representing not only the multidisciplinary areas which have traditionally been part of CLEF but also covering sectors new to CLEF, was established in order to stimulate the submission of innovative papers.

The laboratories continued and expanded on the traditional CLEF track philosophy. Two different forms of labs were offered: benchmarking activities proposing evaluation tasks and comparative performance analyses, and workshop-style labs that explore issues of information access evaluation and related fields. A lab selection committee reviewed the lab proposals and decided on those to accept for CLEF 2010. Criteria for selection included soundness of methodology, feasibility of task, use case, existence of business case/industrial stakeholders, number of potential participants, clear movement along a growth path, scale of experiments, interdisciplinarity, originality, and reusability of results. The objective of this new procedure was twofold: (i) to try to address a long-standing issue in CLEF, i.e., tracks that are never-ending due to their enthusiastic volunteer basis, by ensuring a fair and objective review process; (ii) to try to make the benchmarking activities as adherent as possible to the challenges now emerging for multilingual and multimedia information access system development. The results of the activities of the labs are reported in a separate publication, namely the working notes, distributed during CLEF 2010 and available online. It is hoped that they will also give rise to post-conference publications in journals or separate volumes.

The papers accepted for the conference include research on resources, tools, and methods; experimental collections and datasets, and evaluation methodologies. In addition, two keynote talks will highlight future directions for experimental evaluation from an academic and an industrial point of view: Norbert Fuhr (University of Duisburg-Essen, Germany) will explore "IR between Science and Engineering, and the Role of Experimentation", while Ricardo Baeza-Yates (Yahoo! Research and Universitat Pompeu Fabra, Spain) will talk about "Retrieval Evaluation in Practice".

The conference program was interleaved with overviews of the lab activities and reports from other major worldwide evaluation initiatives: the *Text REtrieval Conference (TREC)* in the USA, the *NII-NACSIS Test Collection for IR Systems (NTCIR)* in Japan, the *INitiative for the Evaluation of XML Retrieval (INEX)* in Australia, the *Russian Information Retrieval Evaluation Seminar (ROMIP)* in Russia, and the *Forum for Information Retrieval Evaluation (FIRE)* in India.

In order to further stimulate the discussion and involve the audience, two panels were organized. The first panel, "The Four Ladies of Experimental Evaluation", saw four key researchers – Donna Harman (National Institute of Standards and Technology, NIST, USA), Noriko Kando (National Institute of Informatics, NII, Japan), Mounia Lalmas (University of Glasgow, UK), and Carol Peters (Institute of Information Science and Technologies, Italian National Research Council, ISTI-CNR, Pisa, Italy) – who have created and run the main retrieval evaluation initiatives of the last two decades, discussing what has been achieved so far and what still needs to be achieved by evaluation campaigns. The second panel, "A PROMISE for Experimental Evaluation", was presented by the partners of the *Participative Research labOratory for Multimedia and Multilingual Information Systems Evaluation (PROMISE)*, an EU FP7 Network of Excellence. PROMISE aims at advancing the experimental evaluation of complex multimedia and multilingual information systems in order to support the decision making process of individuals, commercial entities, and communities who develop, employ, and improve such complex systems.

CLEF 2010 was the first and launching event organized by the PROMISE network. PROMISE envisions the relationship between multilingual and multimedia information access systems and experimental evaluation as a kind of "building" with three levels:

- **Basement**: The conference on the evaluation of multilingual and multimedia information systems, the evaluation methodologies and metrics developed to embody realistic use cases and evaluation tasks, and the techniques proposed to bring more automation to the evaluation process constitute the basement of CLEF and will provide the foundations for promoting and supporting scientific and technological advancements.
- **Pillars**: The regular and thorough evaluation activities carried out in the labs, and the adoption of realistic use cases and evaluation tasks designed to satisfy compelling user and industrial needs, represent the pillars of CLEF. They will stimulate research and development in the multilingual and multimedia information systems field and will contribute to the creation and promotion of a multidisciplinary R&D community, which brings together the expertise needed to develop these complex systems.
- **Roof**: The "basement" and the "pillars" of CLEF will provide the necessary support for the design and development of the next generation multilingual and multimedia information systems needed to address emerging user needs and to cope with the current drive for interaction between content, users, languages, and tasks.

The success of CLEF 2010 would not have been possible without the invaluable contributions of all the members of the Program Committee, Organizing Committee, students, and volunteers that supported the conference in its various stages. Thank you all! In particular, we would like to express our gratitude to the sponsoring organizations for their significant and timely support.

These proceedings have been prepared with the assistance of the Center for the Evaluation of Language and Communication Technologies (CELCT), Trento, Italy, under the coordination of Pamela Forner.

July 2010

Maristella Agosti
Nicola Ferro
Carol Peters
Maarten de Rijke
Alan Smeaton

Organization

CLEF 2010 was organized by the Information Management Systems (IMS) research group of the Department of Information Engineering (DEI) of the University of Padua, Italy.

Honorary Chair

Carol Peters, Institute of Information Science and Technologies (ISTI), Italian National Research Council (CNR), Italy

General Chairs

Maristella Agosti	University of Padua, Italy
Maarten de Rijke	University of Amsterdam, The Netherlands

Program Chairs

Nicola Ferro	University of Padua, Italy
Alan Smeaton	Dublin City University, Ireland

Labs Chairs

Martin Braschler	Zurich University of Applied Sciences, Switzerland
Donna Harman	National Institute of Standards and Technology (NIST), USA

Resource Chair

Khalid Choukri	Evaluations and Language resources Distribution Agency (ELDA), France

Organization Chair

Emanuele Pianta	Center for the Evaluation of Language and Communication Technologies (CELCT), Italy

Organizing Committee

Maria Bernini	DEI administration, University of Padua, Italy
Antonio Camporese	DEI administration, University of Padua, Italy
Marco Dussin	IMS research group, University of Padua, Italy
Pamela Forner	CELCT, Italy
Ivano Masiero	IMS research group, University of Padua, Italy
Sabrina Michelotto	DEI administration, University of Padua, Italy
Giovanni Moretti	CELCT, Italy
Gianmaria Silvello	IMS research group, University of Padua, Italy

Program Committee

Eneko Agirre	University of the Basque Country, Spain
Giambattista Amati	Fondazione Ugo Bordoni, Italy
Julie Berndsen	University College Dublin, Ireland
Pia Borlund	Royal School of Library and Information Science, Denmark
Pavel Braslavski	Yandex and Ural State Tech. University, Russia
Chris Buckley	Sabir Research, USA
Pável Calado	Superior Technical Institute, Portugal
Tiziana Catarci	Sapienza University of Rome, Italy
Stefano Ceri	Politecnico di Milano, Italy
Bruce Croft	University of Massachusetts, Amherst, USA
Franca Debole	ISTI-CNR, Pisa, Italy
J. Stephen Downie	University of Illinois at Urbana-Champaign, USA
Susan Dumais	Microsoft Research, USA
Floriana Esposito	University of Bari, Italy
Marcello Federico	Fondazione Bruno Kessler, Italy
Norbert Fuhr	University of Duisburg, Germany
Fredric Gey	University of California, Berkeley, USA
Marcos André Gonçalves	Federal University of Minas Gerais, Brazil
Julio Gonzalo	National Distance Learning University, Spain
Gregory Grefenstette	Exalead, France
Allan Hanbury	Information Retrieval Facility, Austria
Peter Ingwersen	Royal School of Library and Information Science, Denmark
Perla Innocenti	University of Glasgow, UK

Kalervo Järvelin	University of Tampere, Finland
Gareth Jones	Dublin City University, Ireland
Theodore Kalamboukis	Athens University of Economics and Business, Greece
Noriko Kando	National Institute of Informatics, Japan
Sarantos Kapidakis	Ionian University, Greece
Jussi Karlgren	Swedish Institute of Computer Science, Sweden
Kazuaki Kishida	Keio University, Japan
Mikko Kurimo	Helsinki University of Technology, Finland
Mounia Lalmas	University of Glasgow, UK
Ray Larson	University of California, Berkeley, USA
Chin-Yew Lin	Microsoft Research Asia, China
Saturnino Luz	Trinity College Dublin, Ireland
Bernardo Magnini	Fondazione Bruno Kessler, Italy
Prasenjit Majumder	DAIICT, India
Thomas Mandl	University of Hildesheim, Germany
Paul McNamee	Johns Hopkins University, USA
Mandar Mitra	Indian Statistical Institute, India
Stefano Mizzaro	University of Udine, Italy
Alistair Moffat	University of Melbourne, Australia
Viviane Moreira Orengo	Federal University of Minas Gerais, Brazil
Henning Müller	Univ. of Applied Sciences Western Switzerland, Switzerland
Jian-Yun Nie	University of Montreal, Canada
Douglas W. Oard	University of Maryland, USA
Nicola Orio	University of Padua, Italy
Christos Papatheodorou	Ionian University, Greece
Gabriella Pasi	University of Milan Bicocca, Italy
Anselmo Peñas	National Distance Learning University, Spain
Vivien Petras	Humboldt University, Germany
Jean-Michel Renders	Xerox Research Centre, France
Seamus Ross	University of Toronto, Canada
Ian Ruthven	University of Strathclyde, UK
Tetsuya Sakai	Microsoft Research Asia, China
Diana Santos	SINTEF Information and Communication Technology, Norway
Giuseppe Santucci	Sapienza, University of Rome, Italy
Frederique Segond	Xerox Research Centre, France
Giovanni Semeraro	University of Bari, Italy
Paraic Sheridan	Centre for Next Generation Localisation, Ireland

Ian Soboroff   National Institute of Standards and
                                Technology, USA
Alexander Sychov   Voronezh State University, Russia
John Tait   Information Retrieval Facility, Austria
Letizia Tanca   Politecnico di Milano, Italy
Elaine Toms   Dalhousie University, Canada
Andrew Trotman   University of Otago, New Zealand
Natalia Vassilieva   HP Labs Russia, Russia
Felisa Verdejo   National Distance Learning University, Spain
Ellen Voorhees   National Institute of Standards and
                                Technology, USA
Gerhard Weikum   Max-Planck Institute for Informatics,
                                Germany
Christa Womser-Hacker   University of Hildesheim, Germany
Justin Zobel   University of Melbourne, Australia

Sponsoring Institutions

CLEF 2010 benefited from the support of the following organizations:

Information Retrieval Facility (IRF), Austria
Xerox Research Centre Europe (XRCE), France
Consorzio per la formazione e la ricerca in ingegneria dell'informazione in Padova
    (COFRIDIP), Italy
Department of Information Engineering, University of Padua, Italy
University of Padua, Italy

Table of Contents

Evaluation Methodologies and Metrics (1)

Evaluation Methodologies and Metrics (2)

Panels

IR between Science and Engineering, and the Role of Experimentation

Norbert Fuhr

Department of Computer Science and Applied Cognitive Science,
Faculty of Engineering, University of Duisburg-Essen, 47048 Duisburg, Germany
norbert.fuhr@uni-due.de

Abstract. Evaluation has always played a major role in IR research, as a means for judging about the quality of competing models. Lately, however, we have seen an over-emphasis of experimental results, thus favoring engineering approaches aiming at tuning performance and neglecting other scientific criteria. A recent study investigated the validity of experimental results published at major conferences, showing that for 95% of the papers using standard test collections, the claimed improvements were only relative, and the resulting quality was inferior to that of the top performing systems [AMWZ09].

In this talk, it is claimed that IR is still in its scientific infancy. Despite the extensive efforts in evaluation initiatives, the scientific insights gained are still very limited – partly due to shortcomings in the design of the testbeds. From a general scientific standpoint, using test collections for evaluation only is a waste of resources. Instead, experimentation should be used for hypothesis generation and testing in general, in order to accumulate a better understanding of the retrieval process and to develop a broader theoretic foundation for the field.

Reference

[AMWZ09]   Armstrong, T.G., Moffat, A., Webber, W., Zobel, J.: Improvements that don't add up: ad-hoc retrieval results since 1998. In: Cheung, D.W.-L., Song, I.-Y., Chu, W.W., Hu, X., Lin, J.J. (eds.) CIKM, pp. 601–610. ACM, New York (2009)

M. Agosti et al. (Eds.): CLEF 2010, LNCS 6360, p. 1, 2010.
© Springer-Verlag Berlin Heidelberg 2010

Retrieval Evaluation in Practice

Ricardo Baeza-Yates

Yahoo! Research and Universitat Pompeu Fabra
Barcelona, Spain
rbaeza@acm.org

Abstract. Nowadays, most research on retrieval evaluation is about comparing different systems to determine which is the best one, using a standard document collection and a set of queries with relevance judgements, such as TREC. Retrieval quality baselines are usually also standard, such as BM25. However, in an industrial setting, reality is much harder. First, real Web collections are much larger – billions of documents – and the number of all relevant answers for most queries could be several millions. Second, the baseline is the competition, so you cannot use a weak baseline. Third, good average quality is not enough if, for example, a significant fraction of the answers have quality well below average. On the other hand, search engines have hundreds of million of users and hence click-through data can and should be used for evaluation.

In this invited talk we explore important problems that arise in practice. Some of them are: Which queries are already well answered and which are the difficult queries? Which queries and how many answers per query should be judged by editors? How we can use clicks for retrieval evaluation? What retrieval measure we should use? What is the impact of culture, geography or language in these questions?

All these questions are not trivial and depend in each other, so we only give partial solutions. Hence, the main message to take away is that more research in retrieval evaluation is certainly needed.

M. Agosti et al. (Eds.): CLEF 2010, LNCS 6360, p. 2, 2010.
© Springer-Verlag Berlin Heidelberg 2010

A Dictionary- and Corpus-Independent Statistical Lemmatizer for Information Retrieval in Low Resource Languages

Aki Loponen and Kalervo Järvelin

Department of Information Studies and Interactive Media,
FI-33014 University of Tampere, Finland
{Aki.Loponen,Kalervo.Jarvelin}@uta.fi

Abstract. We present a dictionary- and corpus-independent statistical lemmatizer StaLe that deals with the out-of-vocabulary (OOV) problem of dictionary-based lemmatization by generating candidate lemmas for any inflected word forms. StaLe can be applied with little effort to languages lacking linguistic resources. We show the performance of StaLe both in lemmatization tasks alone and as a component in an IR system using several datasets and query types in four high resource languages. StaLe is competitive, reaching 88-108 % of gold standard performance of a commercial lemmatizer in IR experiments. Despite competitive performance, it is compact, efficient and fast to apply to new languages.

1 Introduction

Word inflection is a significant problem in information retrieval (IR). In monolingual IR, query word inflection causes mismatch problems with the database index. Likewise, in cross-lingual IR (CLIR) inflected query words, tokens, cannot be found as translation dictionary headwords. These challenges plague morphologically complex languages but disturb retrieval also in simpler ones.

The problems of inflection have been addressed using both stemming (e.g. [4], [8], [12], [13]) and lemmatization (e.g. [6], [7]). The potential benefits of lemmatization over stemming, especially in morphologically complex languages, are increased precision due to less ambiguity in all text-based IR and, in CLIR, the support to accurate token translation by directly matching dictionary headwords. Lemmatizers traditionally use morphological rules and dictionaries [7]. Dictionary-based methods are however powerless when encountering out-of-vocabulary (OOV) words. OOV words are significant in all IR, because they often are specific in representing the information needs behind short queries [6, p. 31]. The existing contemporary approaches to dictionary independent lemmatization are based on supervised (e.g. [10], [17]) and unsupervised (e.g. [19]) machine learning techniques. Supervised techniques need a training corpus involving a large set of morphologically analyzed tokens. The disadvantage of this method is that the preparation of the training corpus is time-consuming, in particular when the tokens are tagged manually. The main limitation of unsupervised learning techniques is their general and corpus dependent nature that decreases their performance in specific tasks such as lemmatization.

M. Agosti et al. (Eds.): CLEF 2010, LNCS 6360, pp. 3–14, 2010.

In this paper, we present a statistical corpus- and dictionary-independent lemmatizer for IR, *StaLe*, that effectively deals with all words encountered in texts and queries, OOV words included, and that does not need any tagged corpora. StaLe is based on statistical rules created from a relatively small corpus of token-lemma pairs where each token is an inflected or derived form of a lemma. The method is a simplification of the transliteration rule based translation method for cross-language OOV words by Pirkola and colleagues [16]. StaLe generates candidate lemmas for any word form and consonant gradation. The method can be applied with little effort to languages lacking linguistic resources, because it is dictionary-independent.

Being statistical, our lemmatizer can only generate high quality lemmas through statistical features of its rules, and language specific parameters. Therefore the lemmatizer can generate noisy results, i.e. nonsense lemmas not belonging to the language. By parameter setting, StaLe can be made to emphasize either *lemmatization precision* or *lemmatization recall*.

We believe that lemmatization recall has priority because a human user can often recognize correct looking lemmas among nonsense. Also in automatic IR, when the database index and the queries are statistically lemmatized, we hypothesize that nonsense lemmas do not significantly deprave effectiveness because such lemmas in the database index do not frequently match nonsense query lemmas unless they originate from the same word form.

We show the performance of StaLe both in plain lemmatization tasks and as a component in an IR system using several datasets. For the lemmatization tests, we use the CLEF 2003 and PAROLE [14] collections to create the test word lists and a dictionary-based commercial lemmatizer TWOL [7] as the gold standard. It is as well-known and effective lemmatizer with large internal dictionaries for all the languages used in testing. We also experiment whether the gold standard can be improved when StaLe is used as an additional resource.

For the IR tests, we use CLEF 2003 full-text collections and the retrieval tool-kit Lemur, version 4.7 [9]. We couple with Lemur both StaLe and two state-of-the-art baselines, a lemmatizer and a stemmer. In the tests, we employ Finnish, Swedish, German and English, which represent morphological variation from highly complex to very simple. The main contribution of the present paper is to show the light-weight lemmatizer, StaLe, as an effective lemmatizer for low resource languages. This is why we test StaLe against standard techniques in high resource languages: only these languages allow using strong baselines. As the proposed method StaLe is shown effective in handling this set of languages, we can trust that it is effective in handling other languages, e.g. ones with scarce resources.

The paper is structured as follows: in Chapter 2 we describe the statistical lemmatization method and discuss its working principles. We then present in Chapter 3 two kinds of test situations to measure the effectiveness of StaLe, and also describe how StaLe was set up for these tests. The results of the experiments are presented in Chapter 4 and discussed in Chapter 5 where we also make concluding remarks.

2   The StaLe Lemmatization Method

Our goal was to create a flexible and light, purely statistical lemmatizer, StaLe, which could operate with OOV words as well as with common vocabulary, and also be easily adaptable to new languages and domains. The lemmatizer was primarily aimed for languages with little resources for IR. Research and development in IR and practical retrieval in those languages would benefit from such tools.

StaLe produces the *result lemmas* for given *input token* by applying suitable rules from a *rule set* on the input token to create *candidate* lemmas. The candidate lemmas are then sorted by their confidence factor values and then pruned according to *parameter values* in the *candidate check-up phase*. The lemmas that qualify the check-up phase are the *result lemmas*. The parameters may also allow the input token itself to be added as a result lemma, because it may, with some probability, already be an ambiguous lemma. With a pre-generated rule database and trained parameter values the actual process of StaLe is quite simple as shown on the left side in Fig. 1.

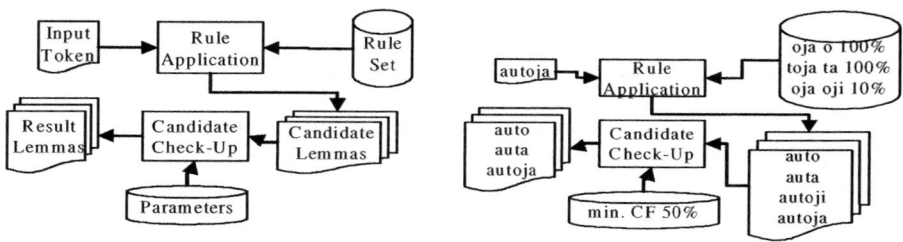

Fig. 1. The StaLe lemmatization process and an example process with input word "autoja"

StaLe gives always at least one result word for input token. The input token itself is given as a result when no other candidate lemma can be created. An example how an input token is processed is given on the right side in Fig. 1. There an input token "autoja" (Finnish for "cars", plural of partitive case) is the input. Three rules are found and applied to create the candidate lemma list including the input token itself. For example, the first rule suggest by 100 % confidence that the ending 'oja' should be replaced by 'o'. After the candidate check-up the lemma with lower confidence factor than defined in the parameters is dropped from the result list.

Our method is meant to serve primarily as a pure lemmatizer. *Compound words* are processed as *atomic* and not split, because corpus and dictionary independent compound splitting is highly ineffective. Airio [1] also showed that decompounding in monolingual IR is not vital.

Because StaLe is statistical and dictionary-independent, it may sometimes create nonsense words in addition to correct lemma forms, and therefore lemmatization precision is poorer than when using dictionary-based methods. However, this can be tolerated in IR applications: finding the correct lemmas is primary and possible noise gets sorted out in other parts of IR.

2.1 Rules

The rules are a variation of *TRT-rules* by Pirkola and colleagues [16] with no restrictions on the length of the rule. A rule consists of *source* and *target strings*, the *position of the rule* and *statistic values* of the rule.

A rule string for English could, for example, be "fier fy e 8 100.0". The source string "fier" is the part of the unprocessed token that is replaced with the target string "fy" to form the candidate lemma. The rule also has a position value "e", which defines whether the rule applies in the beginning (b), in the end (e) or in the middle (m) of the given token. Rules can be generated for prefixes and suffixes.

Both strings in a rule share a *context character* that serves as a binding between the stem and affix of the word. The context character helps to prune the rule list and also separates rules dealing with consonant gradation from non-gradational rules.

Rules have two numerical values identical to the original TRT-rules. The first one is *rule frequency* representing the relative occurrence of the rule in the rule training list (in the example the rule frequency is the value "8"). The second value is *rule confidence factor* which measures how common the rule transformation is among all the transformations that match the same source string (value "100.0" in the example).

2.2   The Rule Training Lists and Parameter Tuning

A paired list of inflected tokens and their corresponding lemmas are needed to create the rules. The tokens and lemmas in the list should be representative for the language into which the method is applied and should be extracted from real texts. However, the list does not need to be an exhaustive representation of the language; quite the contrary: a small sample is enough to represent the most common cases.

Rules could be constructed intellectually or later hand-picked into the rule-set, but then they would lack first-hand information about the distribution and cross-relations between the rules and would lead into too severe deterioration of lemmatization precision with input data from real life materials. The statistical values are necessary for making distinctions between rules when processing words.

The size of the training list is relative to the morphological complexity of the target language [15]: the training list should be larger for languages with lots of morphological variation than for "simpler" languages. Usually, however, some morphological features dominate while others are marginal and speculative, thus making it possible to gain good results without utilizing all possible rules [5]. Therefore it is possible to create a good set of rules using relatively small training list.

A key aspect for the applicability of StaLe is the one-time set-up effort it needs in the form of the training list. Our training lists for complex languages contained roughly 10.000 to 30.000 training pairs, so this effort varies from a few days to two weeks of routine work as estimated by Lindén [10]; much less than needed for coding a new stemmer or a dictionary-based lemmatizer.

Lemmatization parameters reduce noise and improve lemmatization precision by defining a minimum rule confidence factor value for applied rules and controlling the number of created lemma candidates with an upper limit. Parameters are also needed to cope with situations where the input token already is a lemma and therefore should not be lemmatized, or the input token is not in the domain of the rules (i.e. if the rules are only for nouns and the input token is a verb). The parameters are strongly related to the rule sets and therefore should be trained using the same rules that are used in actual lemmatization. If a rule list is modified then it is appropriate to train a new set of parameters. The training is a matter of minor experiments.

3 Experiments

To assess the effectiveness of StaLe we did three kinds of performance tests. First we wanted to see how good the method is in plain lemmatization. This was done by processing paired token-lemma -wordlists with a program, which used StaLe to turn the tokens to candidate lemmas and compared those to the given lemmas.

Secondly we wanted to analyze how StaLe performs in IR when the document database index and the search topics are processed with StaLe. The results of these tests were compared against results achieved with a dictionary based gold standard lemmatizer and a stemmer in four languages.

Thirdly, we experimented with the significance of verbs in information retrieval tests. We examined typical CLEF 2003 query titles and descriptions and found that only about 5 % of proper query words (i.e. topic words with stop-words pruned) are verbs. Bendersky and Croft [3] showed that specifically for long queries noun phrases are more effective that verb phrases. Nouns also carry most of the semantics in a query and represent the actual "things" that are retrieved [2], [11]. Therefore we assumed that the lack of verbs processing has at most a neglible effect on IR performance and to support our estimation we ran parallel IR tests with queries with verbs intact and with queries which had the verbs intellectually identified by their grammatical role in sentences and removed.

We used four languages, Finnish, Swedish, German and English, in all tests. For each language we trained rule sets and parameters which were used in all tests in each language. For all languages the input token itself was also included in the result list before the check-up phase to compete for listing in the final results.

3.1 Rule Lists

One rule set for each language was generated from parts of CLEF 2003 collections. Only nouns, adjectives and pronouns were used: verbs were excluded because the inflection of verbs differs significantly from the other word classes and would therefore increase the number of rules and excess noise.

For each of the four languages we varied the training list sizes to approximate the smallest set of rules that still would produce good results. The selected rule training list and corresponding rule list sizes are shown in Table 1. The estimated amount of 30 000 word pairs was enough to produce a good rule set for Finnish, and the other languages needed much less to reach good enough rule sets.

The sizes of the trained rule lists reflect the differences in the number of morphosyntactic features in various languages and the actual frequency of different inflectional word forms in rule training lists. English has only two features in grammatical case and German four, which explain the small size of the rule list. [15]

Table 1. Rule training list sizes and corresponding rule set sizes

language	training word pairs	number of rules	language	training word pairs	number of rules
Finnish	28610	5371	German	16086	426
Swedish	14384	960	English	9447	242

Parameters for each language were trained by testing with training lists generated with TWOL lemmatizers. The input token confidence factor values were selected from [6] and [15]: they are not related to any individual rule list, but to the language that the parameters and rule lists are used in, and do not therefore need training.

3.2 Lemmatization Tests

With lemmatization experiments we show how well StaLe lemmatizes inflected words and how well it treats words already in lemma form or words belonging to word classes that are outside the lemmatization rules. We used language specific versions of TWOL as the gold standard method and also a naïve baseline. In addition to testing StaLe and TWOL individually, we tested a combination in the "mixed" test setting (see below) where we lemmatized tokens with TWOL and treated the OOV tokens with StaLe. This is expected to maximize lemmatization performance.

For each language, we created three separate test lists from the CLEF collection for Finnish, PAROLE collection for Swedish and CLEF collections for German and English. The three lists for each language contained token-lemma pairs where the token was obtained from the collection and the lemma was either processed with TWOL or intellectually formed. To equalize the test situation for both lemmatizers, each list included only nouns, pronouns and adjectives. The number of words and the percentage of inflected words in the test lists are presented in Table 2.

Table 2. The number of words and the percentage of inflected words in the test lists

	Finnish	Swedish	German	English
twolled	13825 (79%)	15942 (60%)	7834 (49%)	1229 (19%)
mixed	4792 (76%)	1219 (52%)	3894 (21%)	2263 (1%)
OOV	373 (100%)	144 (99%)	588 (32%)	1084 (79%)

Firstly, we formed a large test list using TWOL so that we could do a straight comparison between StaLe and the gold standard. This test list was created by extracting words recognized by TWOL in each language from text collections and identifying the lemmas by TWOL.

Secondly we created a mixed test list by lemmatizing the words with TWOL and then intellectually finding lemmas for words that were OOV for TWOLs. This test list modeled a situation, where foreign words, unrecognized proper names, spelling errors and ad hoc words are present. The percentages of OOV-words in the mixed test list for Finnish are 22 %, for Swedish 23 %, for German 10 % and for English 37 %.

The third list contained only OOV words from the same text collections. The words were the words left unlemmatized by TWOL in the process of forming the first test lists. The test lists were finalized by intellectually processing the OOV material so that only inflected words were picked, because we wanted to analyze how well StaLe lemmatizes inflected words that are out of dictionary-based method's vocabulary.

3.3 IR Tests

IR tests were conducted to compare StaLe to the state-of-the-art baselines, and to investigate our assumptions that lemmatization recall has priority over lemmatization precision in an IR situation and that verbs are not essential for good performance.

We used CLEF 2003 full-text collections for each of the four languages and the query topics were also from CLEF 2003: 45 topics for Finnish, 54 for Swedish and English, and 56 for German. The IR system used to create the search indexes and to perform the retrieval operations was Lemur-toolkit version 4.7 into which the stemmer and both lemmatization methods were incorporated.

For each language, we built four indexes for the document collections: one with StaLe and the other three with the inflected baseline (marked *Bline*), the stemmer and the gold standard, respectively. The inflected baseline index was built without any morphological processing using the text tokens as they appeared. The stemmed index was created using the Snowball stemmers [18] for each language. Language specific versions of TWOL were used to return the gold standard lemmas. StaLe used the same rules and parameters as in the lemmatization tests. All the words in the documents were processed equally with the StaLe and thus also verbs, adverbs and particles were processed as nouns to evaluate a blind application of StaLe.

Three types of queries were formed for each language. *Long queries* included the title and description fields of the query topics. *Long queries w/o verbs* were the same as the long queries, but with verbs intellectually removed prior to morphological processing. *Short queries* included only the title fields of the topics. Each query of each of the three types was processed with the four methods and then matched only to indexes created with the same method.

3.4 Methods of Analysis

To assess the effectiveness of lemmatization we calculate lemmatization recall, lemmatization precision, F_2 values and mean values of those for each input token list. Lemmatization recall is the number of correct lemmas received with the input tokens inserted among the generated candidate lemmas. Lemmatization precision measures the proportion of false candidates in the result list.

The F_2-measure is a recall-biased derivate of the F_n-measure. F-measure combines recall and precision into one metric so that the results are easier to judge. The bias towards recall was made because lemmas not belonging to the language do not matter much in IR and are rather easy to distinguish in manual processes. For example, if an inflected word is translated into another language, the extra noise does probably not matter much because the translation dictionary eliminates the "non-words".

In the IR tests we measure effectiveness with *mean average precision* (MAP) and *precision at ten retrieved documents* (P@10). Statistical significance between the results of different methods was analyzed using *paired two-tailed t-test*. The standard significance level of 95 % was selected as the limit for statistical significance.

4   Results

4.1   Lemmatization Tests

Table 3 gives the F_2 results for the lemmatization tests. The joint method of TWOL and StaLe is marked "T+S". The results between the baseline (Bline), on the one hand, and the two lemmatization methods, on the other, are all statistically significant

Table 3. The F_2-results of the lemmatization tests

language	Finnish			Swedish			German			English		
test list	*twolled*	*mixed*	*OOV*	*twolled*	*mixed*	*OOV*	*twolled*	*mixed*	*OOV*	*twolled*	*mixed*	*OOV*
T+S	-	93.75	-	-	95.45	-	-	95.26	-	-	98.50	-
TWOL	100	86.44	1.71	100	91.84	1.08	100	84.03	0.00	100	99.23	12.33
StaLe	67.74	68.73	58.19	76.48	78.18	67.01	90.72	88.62	74.55	91.09	95.82	84.45
Bline	24.51	23.85	0.25	42.10	48.54	0.39	70.46	60.95	0.00	81.10	98.88	12.33

and therefore the statistical significance is not indicated in Table 3. Note that for the gold standard (TWOL) the "twolled" word list necessarily yields F2 of 100 %.

In the TWOL-lemmatized test setting StaLe easily exceeds the baseline performance and reaches relatively close to the gold standard. In both Finnish and Swedish the test words were ambiguous and usually several suitable rules for each test word were found. The F_2 for Finnish is 67.74 % and for Swedish 76.48 %. However, the lemmatization recalls are 88.6 % and 96.4 % respectively. With German and English the test list had a larger proportion of uninflected and unambiguous tokens which helped StaLe to reach F_2 scores around 91 %.

With mixed test list TWOL's scores degraded at most 15.97 % units (German) and only 0.77 % units with English. However, TWOL returned the test word itself if no lemma could be found. When the baseline gave 98.88 % for English, it is clear that the test list had only few inflected OOVs. Because StaLe created noisy results, it had a score below the baseline. Overall, StaLe's scores were stable across the test lists and for three languages StaLe was 20 to 74 % units above the baseline.

In Finnish the combination method improved the results by 7.31 % units over plain TWOL. An improvement of 3.61 % units was gained in for Swedish and 11.23 % units in German. In English the results deteriorated 0.73 % units. The inflected OOV word list naturally was nearly impossible for TWOL and baseline. However, StaLe was able to maintain its high level of performance.

4.2 IR Tests

The results for the IR tests are shown in Table 4, where the methods are sorted by their mean average precision (MAP) values. An asterisk indicates statistical significance in comparison with StaLe ($p < 0.05$). If the statistically significant value is smaller than the corresponding value of StaLe, then StaLe performed better in that test case, but if the statistically significant value is larger, then StaLe was inferior.

The results show that morphological processing has a strong effect on IR in Finnish improving MAP about 20 % units from the baseline. StaLe and TWOL received quite similar results with no significant statistical difference: the scores of StaLe were from 92 % to 105 % of the TWOL scores. When verbs were included in long queries TWOL attained better MAP value, but when verbs were removed then StaLe was the winner. With short queries there was no significant difference between StaLe and TWOL. Stemming was nearly as effective as lemmatization in Finnish and the difference against StaLe was insignificant except in long queries without verbs.

In Swedish StaLe was at its weakest in the long queries with verbs, where the difference to TWOL was statistically significant. However, when the verbs were

Table 4. The IR test results

Long queries

Finnish (N=45)			Swedish (N=54)			German (N=56)			English (N=54)		
method	MAP	P@10	method	MAP	P@10	method	MAP	P@10	method	MAP	P@10
TWOL	52.76	35.11	TWOL	42.72*	33.70	TWOL	45.04*	48.57	StaLe	50.78	36.11
StaLe	48.69	33.11	Stem	40.48	32.22	Stem	42.70	48.39	Stem	48.06	35.74
Stem	46.33	30.89	StaLe	39.08	31.30	StaLe	41.40	46.61	TWOL	46.85*	34.44
Bline	31.49*	24.22*	Bline	35.01*	28.70	Bline	38.16	44.64	Bline	44.67*	34.63

Long queries w/o verbs

Finnish (N=45)			Swedish (N=54)			German (N=56)			English (N=54)		
method	MAP	P@10	method	MAP	P@10	method	MAP	P@10	method	MAP	P@10
StaLe	53.84	33.11	TWOL	43.07	33.33	TWOL	44.71*	49.82	StaLe	48.38	36.11
TWOL	51.42	34.00	StaLe	42.40	33.52	Stem	42.49	48.75	Stem	47.25	35.19
Stem	47.89*	30.44*	Stem	41.11	31.85	StaLe	41.67	47.68	TWOL	45.58	35.00
Bline	38.95*	26.22*	Bline	37.11*	29.26*	Bline	37.86*	45.00	Bline	44.18*	33.70

Short queries

Finnish (N=45)			Swedish (N=54)			German (N=56)			English (N=54)		
method	MAP	P@10	method	MAP	P@10	method	MAP	P@10	method	MAP	P@10
TWOL	45.43	29.78	TWOL	38.16	30.37	TWOL	35.56*	44.64*	StaLe	43.47	32.22
StaLe	45.42	31.11	StaLe	37.05	29.07	Stem	32.69	40.36	TWOL	42.52*	30.74
Stem	39.37	27.27	Stem	36.12	28.52	StaLe	31.13	40.00	Stem	42.44	30.56
Bline	30.53*	23.47*	Bline	29.32*	25.19*	Bline	28.11*	36.96*	Bline	42.08	29.63

removed StaLe bypassed stemming and nearly caught TWOL narrowing the difference to insignificant. In short queries the differences were also insignificant.

For German the morphological processing had similar effect as in Swedish. This time, however, StaLe could only reach roughly from 88 % to 93 % of the scores of TWOL. The difference between StaLe and stemming was statistically insignificant. In German the inflected baseline was unaffected when verbs were removed.

The difference between the baseline and the gold standard was the smallest in English. Stemming was marginally better than TWOL, but StaLe was clearly the best method in short and long queries with verbs. Overall, StaLe's scores were from 102 % to 108 % of the gold standard scores. Unlike in other languages, StaLe's MAP results decreased when verbs were removed.

The exclusion of verbs did not seem to have a significant negative effect. On the contrary, the baseline MAPs improved in Finnish (by 7.46 % units) and in Swedish (by 2.1 % units) while the scores of morphological processing also improved with one exception: TWOL fared worse in Finnish. For German and English the scores diminished slightly when the verbs were excluded. Aside from the Finnish baseline, the differences between queries with and without verbs were statistically insignificant.

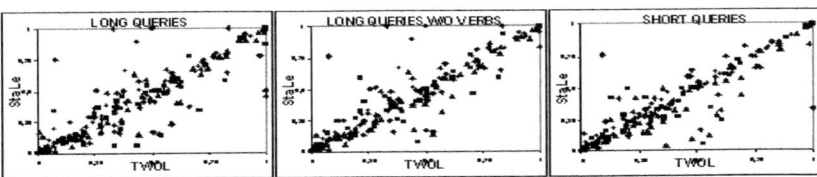

Fig. 2. Topic-by-topic differences between TWOL and StaLe

Figure 2 illustrates the topic-by-topic difference between StaLe and TWOL. These two were chosen as they are the novel method and the gold standard, stemming winning StaLe only in 4/12 MAP cells, and TWOL in 2/12 MAP cells, of Table 4. In Fig. 2, each plot compares the MAP by StaLe to the MAP of TWOL across all four languages and all 209 topics. While there are deviations for the benefit of either lemmatizer, the data points clearly concentrate around the diagonal. This suggests equally robust performance for both lemmatizers and motivates the smallish differences between them in Table 4.

5   Discussion and Conclusions

We have described and tested a dictionary and corpus independent lemmatization method, StaLe, in lemmatization and information retrieval settings for four languages for which advanced morphological and lexical resources are available. The main findings of the tests indicate that StaLe is competitive with state-of-the-art methods used for comparison. This offers strong evidence for StaLe as an effective tool for low resource languages as well. Among the over 6000 living languages spoken globally, the majority have poor language technology resources and are spoken in low resource communities. This makes StaLe attractive. In more detail the findings are as follows.

The strength of StaLe in the plain lemmatization tests was its robustness when more OOV words were introduced in the test word set. With the mixed test list StaLe still has F_2 below the gold standard lemmatizer because the mixed test list contained mostly common vocabulary which is well-coded in the large dictionaries of the gold standard lemmatizer, TWOL. However, building a comprehensive dictionary for a new language is a great effort.

We also tested whether StaLe can improve the mixed test list results of TWOL. In this setting, TWOL processed all words it could and StaLe processed the leftover OOV words. For languages other than English this procedure proved effective improving results from 3.61 to 11.23 % units. In English the score deteriorated below the baseline because inflected OOV words were very scarce in the test list and in those cases the StaLe gave a noisy result set. Adding StaLe to handle OOV words can clearly benefit the gold standard lemmatizer even in high resource languages.

StaLe had effectiveness equal to the dictionary-based gold standard method in IR tests. Morphologically more difficult languages required more rules in the StaLe rulebase. This increased the number of candidate lemmas and therefore lowered the precision scores somewhat. In English StaLe performed better than the competing methods because, firstly, with 242 rules StaLe was able to generalize the rare inflections and the verbs, and secondly, StaLe was able to process OOV words which usually are loan words and proper names.

In the IR tests the performance of the system using StaLe versus the system using the gold standard were within -4.5 to +4 % units from each other, and also clearly above the baseline. Because StaLe can compete with state-of-the-art technologies in high resource languages, it is attractive to implement it for languages, which do not have such resources available.

Despite of the noise caused by the dictionary-independent approach, StaLe performed up to the dictionary-based gold standard. In plain lemmatization, noise lowered the F_2 score, but the IR tests showed that in practical applications noisy lemmas only have a negligible effect.

Our estimation on the negligible effect of the lack of verb processing proved correct: firstly, the removal of verbs from the IR queries did not have significant effects on performance, and performance was not significantly affected when verbs were included and processed as if being nouns. Note, that all document text tokens, including verbs, were indexed by StaLe as if being nouns. From this we conclude that verbs in general have no significant effect on IR and therefore they can be ignored when creating lemmatizers for document retrieval IR systems.

StaLe is easy to implement to new languages and it is robust enough to be effective with relatively little effort. A good rule-set requires at most only about 30 000 token-lemma pairs for morphologically complex language, which takes only a couple of weeks to construct, and less for easier languages. In addition to this, a similar but much smaller word list is also required for parameter training, but basically nothing else needs to be done to get StaLe working in a new language.

While StaLe is a pure lemmatizer it is possible to adjust it to give a probabilistic analysis of the input words because the rules already have values of a probabilistic distribution over grammatical features (number, case, etc.). This could also lead to probabilistic surface syntax analysis. An additional dictionary structure could also be implemented to prune the nonsense lemmas. These belong to further work.

With F_2 score level near 70 % for a morphologically complex language and around 90 % for simpler ones in lemmatization tasks while also performing very well in retrieval tasks, StaLe is a light, robust and effective method for lemmatization.

Acknowledgements. The authors gratefully acknowledge funding from the Academy of Finland under grants #1209960 and #204978.

References

1. Airio, E.: Word normalization and decompounding in mono- and bilingual IR. Information Retrieval 9(3), 249–271 (2006)
2. Baeza-Yates, R., Ribeiro-Neto, B.: Modern Information Retrieval. ACM Press/Addison-Wesley (1999)
3. Bendersky, M., Croft, W.B.: Analysis of Long Queries in a Large Scale Search Log. In: Proceedings of the 2009 workshop on Web Search Click Data, Barcelona, Spain, pp. 8–14 (2009)
4. Frakes, W.B., Baeza-Yates, R.: Information Retrieval: Data Structures & Algorithms. Prentice Hall, Englewood Cliffs (1992)
5. Kettunen, K., Airio, E.: Is a morphologically complex language really that complex in full-text retrieval? In: Salakoski, T., Ginter, F., Pyysalo, S., Pahikkala, T. (eds.) FinTAL 2006. LNCS (LNAI), vol. 4139, pp. 411–422. Springer, Heidelberg (2006)
6. Kettunen, K.: Reductive and Generative Approaches to Morphological Variation of Keywords in Monolingual Information Retrieval. Acta Universitatis Tamperensis 1261. University of Tampere, Tampere (2007)
7. Koskenniemi, K.: Two-level Morphology: A General Computational Model for Word-Form Recognition and Production. Ph.D. Thesis, University of Helsinki, Department of General Linguistics, Helsinki (1983)

8. Krovetz, R.: Viewing morphology as an inference process. In: Proceedings of the 16th ACM/SIGIR Conference on Research and Development in Information Retrieval, Pittsburgh, Pennsylvania, USA, pp. 191–202 (1993)
9. The Lemur Tool-kit for Language Modelling and Information Retrieval, http://www.lemurproject.org/ (visited 30.3.2010)
10. Lindén, K.: A Probabilistic Model for Guessing Base Forms of New Words by Analogy. In: Gelbukh, A. (ed.) CICLing 2008. LNCS, vol. 4919, pp. 106–116. Springer, Heidelberg (2008)
11. Losee, R.M.: Is 1 Noun Worth 2 Adjectives? Measuring Relative Feature Utility. Information Processing and Management 42(5), 1248–1259 (2006)
12. Lovins, J.B.: Development of a Stemming Algorithm. Mechanical Translation and Computation Linguistics 11(1), 23–31 (1968)
13. Majumder, P., Mitra, M., Parui, S.K., Kole, G., Mitra, P., Datta, K.: YASS: Yet Another Suffix Stripper. ACM Transactions on Information Systems (TOIS) 25(4) (2007)
14. Parole, Språkbanken, most common PAROLE words, http://spraakbanken.gu.se/eng/ (visited 30.3.2010)
15. Pirkola, A.: Morphological Typology of Languages for IR. Journal of Documentation 57(3), 330–348 (2001)
16. Pirkola, A., Toivonen, J., Keskustalo, H., Visala, K., Järvelin, K.: Fuzzy translation of cross-lingual spelling variants. In: Proceedings of the 26th ACM SIGIR Conference, Toronto, Canada, pp. 345–353 (2003)
17. Plisson, J., Lavrac, N., Mladenic, D.: A rule based approach to word lemmatization. In: Proceedings of the 7th International Multi-Conference Information Society IS 2004, pp. 83–86 (2004)
18. Snowball, http://snowball.tartarus.org/ (visited 30.3.2010)
19. Wicentowski, R.: Modelling and Learning Multilingual Inflectional Morphology in a Minimally Supervised Framework. Ph.D. Thesis, Baltimore, Maryland, USA (2002)

A New Approach for Cross-Language Plagiarism Analysis

Rafael Corezola Pereira, Viviane P. Moreira, and Renata Galante

Instituto de Informática – Universidade Federal do Rio Grande do Sul (UFRGS)
Caixa Postal 15.064 – 91.501-970 – Porto Alegre – RS – Brazil
{rcpereira,viviane,galante}@inf.ufrgs.br

Abstract. This paper presents a new method for Cross-Language Plagiarism Analysis. Our task is to detect the plagiarized passages in the suspicious documents and their corresponding fragments in the source documents. We propose a plagiarism detection method composed by five main phases: language normalization, retrieval of candidate documents, classifier training, plagiarism analysis, and post-processing. To evaluate our method, we created a corpus containing artificial plagiarism offenses. Two different experiments were conducted; the first one considers only monolingual plagiarism cases, while the second one considers only cross-language plagiarism cases. The results showed that the cross-language experiment achieved 86% of the performance of the monolingual baseline. We also analyzed how the plagiarized text length affects the overall performance of the method. This analysis showed that our method achieved better results with medium and large plagiarized passages.

1 Introduction

Plagiarism is one of the most serious forms of academic misconduct. It is defined as "the use of another person's written work without acknowledging the source". According to Maurer et al. [16], there are several types of plagiarism. It can range from simply copying another's work word-for-word to paraphrasing the text in order to disguise the offense.

A study by McCabe [17] with over 80,000 students in the US and Canada found that 36% of undergraduate students and 24% of graduate students admit to have copied or paraphrased sentences from the Internet without referencing them. Amongst the several methods for plagiarism commonly in practice, Maurer et al. [16] mention cross-language content translation. The authors also surveyed plagiarism detection systems and found that none of the available tools support search for cross-language plagiarism. The increasing availability of textual content in many languages, and the evolution of automatic translation can potentially make this type of plagiarism more common. Cross-language plagiarism can be seen as an advanced form of paraphrasing since every single word might have been replaced by a synonym (in the other language). Furthermore, word order might have changed. These facts make cross-language plagiarism harder to detect.

Cross-language plagiarism, as acknowledged by Roig [28], can also involve self-plagiarism, i.e., the act of translating self published work without referencing the

M. Agosti et al. (Eds.): CLEF 2010, LNCS 6360, pp. 15–26, 2010.
© Springer-Verlag Berlin Heidelberg 2010

original. This offense usually aims at increasing the number of publications. As stated by Lathrop & Foss [13], another common scenario of cross-language plagiarism happens when a student downloads a paper, translates it using an automatic translation tool, corrects some translation errors and presents it as their own work.

The aim of this paper is to propose and evaluate a new method for Cross-Language Plagiarism Analysis (CLPA). The main difference of our method when compared to the existing ones is that we applied a classification algorithm to build a model that can distinguish between a plagiarized and a non-plagiarized text passage. Note that there are two different areas of plagiarism analysis. One area, known as extrinsic plagiarism analysis, uses a reference collection to find the plagiarized passages. The other area, known as intrinsic plagiarism analysis, tries to detect plagiarism without a reference collection, usually by considering differences in the writing style of the suspicious document [15, 30]. In this paper, we focus on extrinsic plagiarism detection. Our task is to detect the plagiarized passages in the suspicious documents (i.e., the documents to be investigated) and their corresponding text fragments in the source documents (i.e., in the reference collection) even if the documents are in different languages. The proposed method is divided into five main phases: language normalization, retrieval of candidate documents, classifier training, plagiarism analysis, and result postprocessing. Since our method aims at detecting plagiarism between documents written in different languages, we used an automatic translation tool to translate the suspicious and the source documents into a single common language in order to analyze them in a uniform way. After the normalization phase, we used a classification algorithm to build a model to enable the method to learn how to distinguish between a plagiarized and a non-plagiarized text passage. To accomplish this task, we selected a pre-defined set of features to be considered during the training phase.

We used an information retrieval system to retrieve, based on the text passages extracted from the suspicious documents, the documents that are more likely to be the source of plagiarism offenses. The idea behind the retrieval phase is that it would not be feasible to perform a detailed analysis between the suspicious document and the entire reference collection. Only after retrieving a small subset of the reference collection, the plagiarism analysis is performed. Finally, detection results were postprocessed to join contiguous plagiarized passages.

In the absence of a corpus with real world plagiarism cases (and considering the work necessary to assemble one), we decided to create an artificial plagiarism corpus called ECLaPa (Europarl Cross-Language Plagiarism Analysis). The corpus is based on the Europarl Parallel Corpus [11], which is a collection of documents generated from the proceedings of the European Parliament. We conducted two different experiments; the first one considers only monolingual plagiarism cases, while the second one considers only cross-language plagiarism cases. The results showed that the cross-language experiment achieved 86% of the performance of the monolingual baseline. The main contribution of this paper is the definition of a new CLPA method as well as its evaluation against an artificial plagiarism collection, which is available to other researchers.

The remainder of this paper is organized as follows: Section 2 discusses related work. Section 3 presents our proposed method for CLPA. Section 4 presents the experiments that were carried out to validate our approach. Section 5 summarizes our contributions and presents the conclusions.

2   Related Work

2.1   Monolingual Plagiarism Detection

Research on document processing has recently devoted more attention to the problem of detecting plagiarism. The standard method for monolingual plagiarism analysis involves comparing chunks from suspicious and source documents. The most popular approach, according to Stein & Eissen [29], is to use the MD5 hashing algorithm [27] to calculate a hash signature (called fingerprint) for each chunk. Identical chunks will have the same fingerprint. Since plagiarized texts are not likely to be identical to its source, the authors proposed a new hashing technique called fuzzy fingerprints [29] to generate the same hash signature for lexically similar chunks. The work by Barrón-Cedeño & Rosso [2] proposes the division of the suspicious documents into sentences, which are then split into word n-grams. The source documents are also split into word n-grams. Then, an exhaustive comparison is performed between the n-grams of each suspicious sentence and the n-grams of each source document. To decide whether the suspicious sentence is plagiarized, the authors applied the containment measure to compare the corresponding sets. The experiments showed that the best results are achieved when using bi-grams (better recall) and tri-grams (better precision).

The method proposed in [8], winner of the PAN 2009 competition [19], computes a matrix of string kernel values in order to find the similarity between suspicious and source documents. An exhaustive pairwise comparison is performed. The 51 top ranked documents are kept for further investigation which involves a pairwise sequence matching technique, called encoplot.

In [10], the authors used overlapping sequences of five words to create an inverted index that maps the 5-word chunk hash value to the list of source documents in which the chunk appears. Once the inverted index is created, each suspicious document is split using the same strategy and the hash values of its 5-word chunks is looked up in the inverted index. In the competition, documents that shared more than 20 chunks were considered similar. After all the common chunks are identified, a merging algorithm is applied to combine the chunks that appear near each other in the suspicious and in the source document. This method achieved the second highest score in the PAN competition.

Methods for monolingual plagiarism detection cannot be directly applied to CLPA because the terms in the suspicious and source text segments will not match. Even if the plagiarized text is an exact translation of the original, word order will change.

2.2   Cross-Language Information Retrieval

CLPA is related to Cross-Language Information Retrieval (CLIR), in which a query in one natural language is matched against documents in another language. The main problem of CLIR is knowing how to map concepts between languages [7], whereas the problem of CLPA is more difficult as it is necessary to match a segment of text in one language to a segment of text of equal content in another language. The sizes of these segments can vary from one sentence to hundreds of pages (i.e., whole books).

There are three traditional approaches for CLIR which are used to bring the query into the language of the documents: (*i*) machine translation (MT) systems; (*ii*) multilingual thesauri or dictionaries; and (*iii*) automatically analyze multilingual corpora to extract term equivalences. CLIR approaches grow in an out of preference throughout the years. While machine readable dictionaries were popular in the late 90's [9], recently MT-based systems are the most employed strategy. For the evaluation campaign CLEF 2009 [20], seven out of ten bilingual approaches used MT systems to bring the queries and the documents into the same language.

2.3   Cross-Language Plagiarism Analysis

Interest on CLPA is recent and it is growing quickly. So far, only a few studies have dealt with CLPA. The work by Barrón-Cedeño et al. [4] relies on a statistical bilingual dictionary created from parallel corpora and on an algorithm for bilingual text alignment. The authors report experiments on a collection composed of 5 original fragments which were used to generate plagiarized versions. The results of the experiments showed that the similarity between the original documents and their plagiarized versions was much higher than the similarity between non-plagiarized documents. MLPlag [5] is a CLPA method based on the analysis of word positions. EuroWordNet is used to transform words into a language independent representation. The authors built two multilingual corpora: JRC-EU and Fairy-tale. The first corpus is composed of 400 randomly selected European Union legislative texts containing 200 reports written in English and the same number of corresponding reports written in Czech. The second corpus represents a smaller set of text documents with a simplified language. This corpus is composed of 54 documents, 27 English and 27 corresponding translations in Czech. The method showed good results. However, the authors stated that the incompleteness of the EuroWordNet may lead to difficulties during cross-language plagiarism detection, especially when handling less common languages.

Other studies propose multilingual retrieval approaches that can help detect document plagiarism across languages. The work by Pouliquen et al. [25] proposes a system that identifies translations and very similar documents among a large number of candidates. The contents of the documents are represented as vectors of terms from a multilingual thesaurus. The similarity measure for documents is the same, independent from the document language. The authors report experiments that search for Spanish and French translations of English documents, using several parallel corpora ranging from 795 to 1130 text pairs and searching in a search space of up to 1640 documents. The result of the experiments showed that the system can detect translations with over 96% precision. The work by Potthast et al. [22] introduces a new multilingual retrieval model (CL-ESA) for the analysis of cross-language similarity. The authors report experiments on a multilingual parallel corpus (JRC-Acquis) and a multilingual comparable corpus (Wikipedia). Recently, in [23] the authors compare CL-ESA to other methods and report that character n-grams achieves a better performance. However, character n-grams will not be suitable for languages with unrelated syntax.

The work by Barrón-Cedeño et al. [3] applies the Kullback-Leibler distance to reduce the number of documents that must be compared against the suspicious document. The main difference to our approach is that they build feature vectors for each reference document and compare these vectors against the vector of the suspicious document. The top ten reference documents with the lowest distance with respect to the vector of the suspicious document are selected to the plagiarism analysis phase.

In 2009, the first PAN competition on plagiarism detection [19] took place. The aim was to provide a common basis for the evaluation of plagiarism detection systems. The corpus released during the competition also contained cross-language plagiarism cases. However, the focus was on monolingual plagiarism and none of the methods used during the competition were designed to detect cross-language plagiarism.

Since this is a new area of research, there is still a lack of resources to enable comparison between techniques. One of the main contributions of this paper is the proposal of a new CLPA method as well as the creation of an artificial cross-language plagiarism corpus. Another contribution is the use of a classification algorithm to create a model that can differentiate plagiarized and non-plagiarized passages. Classification algorithms have been used for intrinsic plagiarism analysis (e.g., [1, 12]), but we do not know of any experiments applying them to the extrinsic plagiarism analysis.

3   The Method

Given a reference corpus D of original documents and a corpus D' of suspicious documents, our proposed method aims at detecting all passages $s \in D'$ which have been plagiarized from a source document $d \in D$. Note that both the original and the suspicious documents can be written in any given language. In order to detect the plagiarized passages, we propose a method divided into five main phases. These phases are depicted in Figure 1 and explained in more detail in the next subsections.

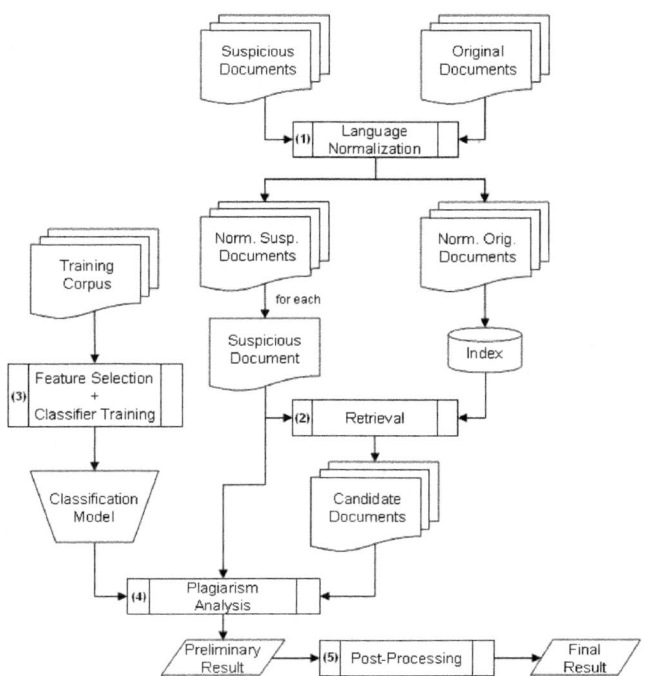

Fig. 1. The five main phases of the proposed method

3.1 Language Normalization

Since our method was designed to handle documents written in several languages, at this phase, the documents are converted into a single common language. We chose English as the default language, since it is the most commonly used language on the Internet and translation resources to and from English are more easily found (e.g., it is easier to find a translator from Finnish to English than from Finnish to Portuguese). Furthermore, according to Koehn [11], English is amongst the easiest languages to translate into as it has fewer inflectional forms than other European languages.

In order to translate the non-English documents into English, an automatic translation tool is used. Before translating, the language in which each document was written is identified using a language guesser.

3.2 Retrieval of Candidate Documents

The main goal of this phase is to use an Information Retrieval (IR) system to select, based on the given suspicious document, the documents in the reference collection that are more likely to be the source of plagiarism offenses. This is an important phase since it would not be feasible to perform a plagiarism analysis between the suspicious document and the entire reference collection. After the candidate documents are retrieved, only a small subset of the collection needs to be analyzed.

Additionally, to reduce the amount of text that needs to be analyzed during the plagiarism analysis phase, we divide the documents into several subdocuments, each one composed of a single paragraph of the source document. The rationale behind this is that after the documents are retrieved, it is not necessary to have a detailed analysis against the entire source documents, only against the subdocuments retrieved. After splitting all the source documents, the reference collection is indexed. Note that from now on, the words passages, paragraphs, and subdocuments are used interchangeably.

Once the reference collection is split and indexed, the system is ready to receive queries to retrieve the candidate subdocuments. The suspicious document is also divided into paragraphs which are used to query the index. Thus, for each text passage (i.e., paragraph) in the suspicious document, we query the index to get the candidate subdocuments with the highest similarity scores. These are the subdocuments with the highest probability of having been used as the source of plagiarism.

Instead of using the entire passage to query the index, we apply a term selection technique and only use the terms in the passage which have an IDF (inverse document frequency) greater than a certain threshold. As a result, the time spent in this phase decreases significantly and the overall result is not affected as the majority of the discarded terms have low discriminating power. Subdocuments that are assigned a low similarity score by the IR system are also discarded.

At the end of this phase, we have a list of at most ten candidate subdocuments for each passage in the suspicious document. Note that these subdocuments might belong to different source documents.

3.3 Feature Selection and Classifier Training

The classification model is based on a set of features. The goal of the classifier is to decide whether a suspicious passage is plagiarized from a candidate subdocument.

Therefore, each time a plagiarism analysis is performed between the suspicious passage and one of the candidate subdocuments, we have two pieces of text to extract information from. The following features are considered:

• *The cosine similarity between the suspicious passage and the candidate subdocument:* The cosine measure discards the order of the terms into consideration. Thus it is useful since the text passages compared may be originated from different languages or may have been obfuscated to confuse plagiarism analysis.

• *The similarity score assigned by the IR system to the candidate subdocument*: the higher the score, the higher the chances of a plagiarism offense.

• *The position of the candidate subdocument in the rank generated by the IR system in response to the suspicious passage used as query*: the candidate subdocuments that are actually plagiarized tend to be at the top of the rank.

• *The length (in characters) of the suspicious and the candidate subdocument:* passages of similar sizes get higher similarity scores. However, even significant differences in length do not necessarily mean plagiarism can be discarded.

Ideally, we should use real cases of plagiarism to train the classifier, but gathering these real examples would not be feasible. Thus, we use an artificial plagiarism corpus. For each document, we are able to identify the location of each plagiarized passage and its respective location in the source document. Based on these annotations we can provide the necessary information in order to train the classifier.

3.4 Plagiarism Analysis

Once the classification model is built, we can submit the test instances to the trained classifier which decides whether the suspicious document s_i has, in fact, plagiarized passages from one or more of the source documents c_j.

3.5 Result Post-Processing

The goal of this phase is to join the contiguous plagiarized passages detected by the method in order to decrease its final granularity score. The granularity score is a measure that assesses whether the plagiarism method reports a plagiarized passage as a whole or as several small plagiarized passages. In order to combine the contiguous detected passages, we use the following heuristics: (*i*) organize the detections in groups divided by the source document; (*ii*) for each group, sort them in order of appearance in the suspicious document; (*iii*) combine adjacent detections that are at most a pre-defined number of characters distant from each other; (*iv*) keep only the detection with the largest length in the source document per plagiarized passage.

4 Experiments

In the next sections we describe the test collection created for the experiments, the metrics employed to evaluate the CLPA method, and the results achieved.

4.1 Test Collection

As mentioned before, due to the lack of resources to enable comparison between different techniques, we decided to create an artificial cross-language plagiarism corpus to evaluate our method. The PAN 2009 Competition corpus was not used in our experiments because it has few cross-language plagiarism cases. We created the corpus based on the Europarl Parallel Corpus [11], which is a collection of parallel documents composed of proceedings of the European Parliament. In this paper, in particular, we used the English-Portuguese and the English-French language-pairs to simulate cross-language plagiarism offenses among these three languages. We named the test collection ECLaPA (Europarl Cross-Language Plagiarism Analysis) and it can be freely downloaded from http://www.inf.ufrgs.br/~viviane/eclapa.html.

The ECLaPA test collection is composed of two corpora, one containing only monolingual plagiarism cases and the other containing only multilingual plagiarism cases. Both corpora contain exactly the same plagiarism cases. In the multilingual corpus the suspicious documents are all written in English, but the source documents are written in Portuguese or in French. Thus, in order to simulate plagiarism cases between these languages, we made a script that randomly selects passages from a Portuguese or French document, locates the equivalent English passages, and inserts them in an English document. To annotate the plagiarized passages, we adopted the format used in PAN [19]. Details on the ECLaPA test collection are shown in Table 1.

Table 1. Characteristics of the Test Collection

---		#Docs	Size	Documents In		
				English	**Portuguese**	**French**
Monolingual	**Suspicious**	300	89MB	300	0	0
Corpus	**Source**	348	102MB	348	0	0
Multilingual	**Suspicious**	300	89MB	300	0	0
Corpus	**Sources**	348	115MB	0	174	174

From the 300 suspicious documents in each corpus, 100 do not contain plagiarism cases. Also, from the 348 source documents in each corpus, 100 were not used as source of plagiarism (in the multilingual corpus, 50 Portuguese documents and 50 French documents). Each corpus has a total of 2169 plagiarism cases, from which about 30% are short passages (less than 1500 characters), 60% are medium passages (from 1501 to 5000 characters), and 10% are large passages (from 5001 to 15000 characters). Each suspicious document can have up to 5 different sources of plagiarism, and from each source, it can have up to 15 plagiarized passages.

4.2 Evaluation Metrics

In order to evaluate the experimental results, the same measures used in the PAN 2009 competition were employed (please refer to [24] for further information on them).

Recall

$$R = \frac{1}{|S|} \sum_{i=1}^{|S|} \left(\frac{\#Detected\_chars\_of\_s_i}{|s_i|} \right)$$

Granularity

$$G = \frac{1}{|S_R|} \sum_{i=1}^{|S_R|} \left(\#Detections\_of\_s_i\_in\_R \right)$$

Precision

$$P = \frac{1}{|R|} \sum_{i=1}^{|R|} \left(\frac{\#Plagiarized\_chars\_of\_r_i}{|r_i|} \right)$$

Overall Score

$$O = \frac{F}{\log_2(1+G)}$$

where s_i is a plagiarized passage from the set S of all plagiarized passages; r_i denotes a detection from the set R of all detections; S_R is the subset of S for which detections exist in R; and F is the harmonic mean of precision and recall.

To be considered correct, a detection r must report at least one correct character in both the plagiarized and the source passage of its corresponding s.

4.3 Method Evaluation

We used the Terrier Information Retrieval Platform [18] as our IR system as it implements several IR techniques enabling us to evaluate our method under different configurations. In particular, we used the TF-IDF weighting scheme as well as stopword removal (a list of 733 words included in the Terrier platform) and stemming (Porter Stemmer [21]). Note that other IR systems could be used as well.

In order to build the classifier, we used Weka [31], which is an open source software that implements several machine learning algorithms. It also provides an API which allowed us to easily integrate it with our solution. Several classification algorithms including BayesNet, J48, NaiveBayes, and AdaBoostM1 were tested. These tests showed that the J48 classification algorithm [26] had the best results. Thus this is the algorithm used in our classifier.

Our task is to detect all the plagiarized passages in the suspicious documents and its corresponding text passages in the source documents. The documents were divided into several subdocuments before the translation in order to keep the original offset and length of each paragraph in the original document. As mentioned in Section 3.1, during the language normalization phase, we have to identify the language of each suspicious and source document in order to translate them all to English. Since only the source documents of the multilingual corpus are written in languages other than English, we applied the language normalization only on this set of documents. We used the LEC Power Translator 12 [14] as our translation tool and the Google Translator [6] as our language guesser. The language guesser correctly classified the language of all source documents in the multilingual corpus.

The next step was to index the collection. We decided to index only the subdocuments with length greater than 250 characters to reduce index size and speed up retrieval. Information about the resulting indexes is shown in Table 2.

Table 2. Information about the indexes

---	Monolingual	Multilingual
# Subdocuments Indexed	134,406	143,861
# Terms	8,379,357	8,786,678
# Unique Terms	37,262	45,370
Size (MB)	38	41

A training collection with the same characteristics as the test collection was created to train the classifier. The training instances were randomly selected from the suspicious documents. For each suspicious document the top ten candidate subdocuments were retrieved. Based on each pair *[suspicious passage, candidate subdocument]*, we can extract the information necessary to create the training instances. The classifier must know if the training instances are positive or negative examples of plagiarism. This information comes from the annotations provided with the corpus.

Since there is no direct way of comparing our CLPA method to the existing ones, we compared the performance of the method when detecting the plagiarism cases of the multilingual corpus against its performance when detecting the plagiarism cases of the monolingual corpus. Since both corpora contain the same plagiarism cases, we believe this experiment can provide us an idea of how well our method handles cross-language plagiarism. The final results of the experiment are shown in Table 3.

Table 3. Experiment Results

---	Monolingual	Multilingual	% of Monolingual
Recall	0.8648	0.6760	78.16%
Precision	0.5515	0.5118	92.80%
F-Measure	0.6735	0.5825	86.48%
Granularity	1.0000	1.0000	100%
Final Score	0.6735	0.5825	86.48%

According to Table 3, the cross-language experiment achieved 86% of the performance (final score) of the monolingual baseline. This is comparable with the state of the art in CLIR. Recall was the most affected measure when dealing with cross-language plagiarism. We attribute the 22% drop in recall to the loss of information incurred by the translation process. As a result, the similarity score assigned by the IR system decreased, leading to more subdocuments being discarded during the retrieval phase. Our post-processing step allowed for perfect granularity in both settings.

To analyze in which situations the method performs better, we investigated to what extent the length of the plagiarized passage affects the results. Table 4 shows the results of the analysis. We divided the plagiarized passages according to their textual length (in characters): short (less than 1500 characters), medium (from 1501 to 5000 characters), and large (from 5001 to 15000 characters).

Table 4. Detailed Analysis

---	Monolingual			Multilingual		
	Short	Medium	Large	Short	Medium	Large
Detected	435	1289	239	242	1190	239
Total	607	1323	239	607	1323	239
%	71	97	100	39	90	100

When considering only the monolingual plagiarism cases, our method detected 90% of the passages (1963). As for the multilingual plagiarism cases, the method detected 77% (1671) of the passages. As expected, the length of the plagiarized passage affects the results considerably. The larger the passage the easier the detection. All large plagiarized passages were detected in both mono and multilingual settings. However, only 40% of short passages in the multilingual corpus were detected. We compared the detection of plagiarism from documents in Portuguese and in French. The results were almost identical (overall score of 0.62 and 0.61 respectively).

5 Conclusions

This paper proposes a new approach to CLPA. The main difference of our method compared to the existing ones is that we used a classification algorithm in order to decide whether a text passage is plagiarized, i.e., instead of defining fixed thresholds we let the classifier learn the most appropriate ones.

To evaluate our method, we created an artificial cross-language plagiarism test collection. The collection is freely available to enable other methods to be compared against ours. In the first experiment, we analyzed the performance of the method when detecting only monolingual plagiarism cases. This experiment was considered the baseline. In the second experiment, we analyzed the performance of the method when detecting cross-language plagiarism cases. The results showed that the cross-language experiment achieved 86% of the performance of the monolingual baseline, which shows that the proposed method is a viable approach for CLPA.

Although we used English, Portuguese, and French in the experiments, our method is designed to be language independent. The necessary resources are an automatic translation tool and an appropriate stemming algorithm to be used in the documents translated into the default language.

As future work, we will test other features to be used during the training phase. We will try to improve detection on small plagiarized passages and test our method with other languages.

Acknowledgements

The authors thank Paul Clough for his helpful comments on an earlier version of this paper. This project was partially supported by CNPq-Brazil and INCT-Web.

References

1. Argamon, S., Levitan, S.: Measuring the Usefulness of Function Words for Authorship Attribution. In: Association for Literary and Linguistic Computing/ Association Computer Humanities (2005)
2. Barrón-Cedeño, A., Rosso, P.: On Automatic Plagiarism Detection Based on n-Grams Comparison. In: Proceedings of the 31st European Conference on IR Research on Advances in Information Retrieval, pp. 696–700 (2009)
3. Barrón-Cedeño, A., Rosso, P., Benedí, J.-M.: Reducing the Plagiarism Detection Search Space on the basis of the Kullback-Leibler Distance. In: Gelbukh, A. (ed.) CICLing 2009. LNCS, vol. 5449, pp. 523–534. Springer, Heidelberg (2009)
4. Barrón-Cedeño, A., Rosso, P., Pinto, D., Juan, A.: On Cross-lingual Plagiarism Analysis using a Statistical Model. In: Proceedings of the ECAI 2008 Workshop on Uncovering Plagiarism, Authorship and Social Software Misuse (2008)
5. Ceska, Z., Toman, M., Jezek, K.: Multilingual Plagiarism Detection. In: Dochev, D., Pistore, M., Traverso, P. (eds.) AIMSA 2008. LNCS (LNAI), vol. 5253, pp. 83–92. Springer, Heidelberg (2008)
6. Google Translator, http://www.google.com/translate_t
7. Grefenstette, G.: Cross-Language Information Retrieval, p. 182. Kluwer Academic Publishers, Boston (1998)
8. Grozea, C., Gehl, C., Popescu, M.: ENCOPLOT: Pairwise Sequence Matching in Linear Time Applied to Plagiarism Detection. In: Proceedings of the SEPLN 2009 Workshop on Uncovering Plagiarism, Authorship and Social Software Misuse, pp. 10–18 (2009)

9. Hull, D.A., Grefenstette, G.: Querying Across Languages, a Dictionary-based approach to Multilingual Information Retrieval. In: 19th Annual International ACM SIGIR Conference on Research and Development in Information Retrieval (1996)
10. Kasprzak, J., Brandejs, M., Křipač, M.: Finding Plagiarism by Evaluating Document Similarities. In: Proceedings of the SEPLN 2009 Workshop on Uncovering Plagiarism, Authorship and Social Software Misuse, pp. 24–28 (2009)
11. Koehn, P.E.: A Parallel Corpus for Statistical Machine Translation. In: MT Summit (2005)
12. Koppel, M., Schler, J.: Authorship Verification as a One-Class Classification Problem. In: Proceedings of the 21st International Conference on Machine Learning. ACM, New York (2004)
13. Lathrop, A., Foss, K.: Student Cheating and Plagiarism in the Internet Era. A Wake-Up Call, p. 255. Libraries Unlimited, Inc., Englewood (2000)
14. LEC Power Translator, http://www.lec.com/power-translator-software.asp
15. Malyutov, M.B.: Authorship Attribution of Texts: A Review in General Theory of Information Transfer and Combinatorics, pp. 362–380. Springer, Heidelberg (2006)
16. Maurer, H., Kappe, F., Zaka, B.: Plagiarism - A Survey. Journal of Universal Computer Science, 1050–1084 (2006)
17. McCabe, D.L.: Cheating among college and university students: A North American perspective. International Journal for Educational Integrity (2005)
18. Ounis, I., Amati, G., Plachouras, V., He, B., Macdonald, C., Johnson, D.: Terrier Information Retrieval Platform. In: Losada, D.E., Fernández-Luna, J.M. (eds.) ECIR 2005. LNCS, vol. 3408, pp. 517–519. Springer, Heidelberg (2005)
19. PAN (2009), http://www.webis.de/pan-09
20. Peters, C., Ferro, N.: CLEF 2009 Ad Hoc Track Overview: TEL & Persian tasks. In: Working Notes of CLEF 2009 (2009)
21. Porter, M.F.: An algorithm for suffix stripping, in Readings in information retrieval, pp. 313–316. Morgan Kaufmann, San Francisco (1997)
22. Potthast, M.: Wikipedia in the pocket: indexing technology for near-duplicate detection and high similarity search. In: Proceedings of the 30th Annual International ACM SIGIR Conference on Research and Development in Information Retrieval, pp. 909–909 (2007)
23. Potthast, M., Barrón-Cedeño, A., Stein, B., Rosso, P.: Cross-language plagiarism detection. In: Language Resources and Evaluation (2010) (Published online on January 30, 2010)
24. Potthast, M., Stein, B., Eiselt, A., Barrón-Cedeño, A., Rosso, P.: Overview of the 1st International Competition on Plagiarism Detection. In: Proceedings of the SEPLN'09 Workshop on Uncovering Plagiarism, Authorship and Social Software Misuse, pp. 1–9 (2009)
25. Pouliquen, B., Steinberger, R., Ignat, C.: Automatic Identification of Document Translations in Large Multilingual Document Collections. In: Proceedings of the International Conference Recent Advances in Natural Language Processing (RANLP 2003), pp. 401–408 (2003)
26. Quinlan, J.R.: C4.5: programs for machine learning, p. 302. Morgan Kaufmann, San Francisco (1993)
27. The md5 message-digest algorithm, http://theory.lcs.mit.edu/~rivest/rfc1321.txt
28. Roig, M.: Avoiding plagiarism, self-plagiarism, and other questionable writing practices: A guide to ethical writing (2010)
29. Stein, B., Eissen, S.M.z.: Near Similarity Search and Plagiarism Analysis. In: From Data and Information Analysis to Knowledge Engineering, pp. 430–437. Springer, Heidelberg (2006)
30. Stein, B., Eissen, S.M.z.: Intrinsic Plagiarism Analysis with Meta Learning. In: SIGIR 2007 - Workshop on Plagiarism Analysis, Authorship Identification, and Near-Duplicate Detection (2007)
31. Weka, http://www.cs.waikato.ac.nz/ml/weka/

Creating a Persian-English Comparable Corpus

Homa Baradaran Hashemi, Azadeh Shakery, and Heshaam Faili

School of Electrical and Computer Engineering,
College of Engineering,
University of Tehran
H.B.Hashemi@ece.ut.ac.ir, {Shakery,HFaili}@ut.ac.ir

Abstract. Multilingual corpora are valuable resources for cross-language information retrieval and are available in many language pairs. However the Persian language does not have rich multilingual resources due to some of its special features and difficulties in constructing the corpora. In this study, we build a Persian-English comparable corpus from two independent news collections: BBC News in English and Hamshahri news in Persian. We use the similarity of the document topics and their publication dates to align the documents in these sets. We tried several alternatives for constructing the comparable corpora and assessed the quality of the corpora using different criteria. Evaluation results show the high quality of the aligned documents and using the Persian-English comparable corpus for extracting translation knowledge seems promising.

1 Introduction and Related Work

The fast growth of the World Wide Web and the availability of information in different languages have attracted much attention in research on cross-language information retrieval (CLIR). One of the main issues in CLIR is where to obtain the translation knowledge [15]. Multilingual corpora, either parallel or comparable, are widely used for this purpose and are available in many language pairs. Comparable corpora are generally obtained from news articles [14,5,22,2], novels [7], available research corpora such as CLEF or TREC collections [23,4,21] or by crawling the web [24,20,26]. On the other hand, parallel corpora are usually obtained from official multilingual documents such as United Nations articles [6] and EU documents [12], multilingual websites [18] or news services [27]. However, Persian, as a widely spoken language in the Middle East, does not have rich resources due to some of its special features and difficulties in constructing corpora [8,1].

The current available Persian corpora are either monolingual and built for special purposes or not big enough for translation purposes. For example, Hamshahri corpus [1] is a monolingual corpus for evaluating Persian information retrieval systems and Bijankhan corpus [3] is a Persian tagged corpus for natural language processing. Current available Persian-English corpora are the Miangah's English-Persian parallel corpus [13] consisting of 4,860,000 words, Tehran English-Persian parallel corpus composed of 612,086 bilingual sentences extracted from movie

M. Agosti et al. (Eds.): CLEF 2010, LNCS 6360, pp. 27–39, 2010.

subtitles in conversational text domain [16], and Karimi's comparable Persian-English corpus [9] consisting of 1100 loosely translated BBC News documents. Since the translation knowledge is usually extracted statistically from the multilingual corpora, the available corpora are not adequate, being either too small or in a special domain.

In this work, we build a Persian-English comparable corpus using Persian articles of Hamshahri newspaper[1] and English articles of BBC News[2]. The total of around 53,000 English documents are aligned with 190,000 Persian documents resulting a comparable corpus of more than 7,500 document pairs. Many studies on exploiting comparable corpora for CLIR assume that comparable corpora are easily obtainable from news articles in different language aligned by date [25,26,2]. Although this assumption may be true in some languages, this is not the case in Persian. News from the news agencies that produce daily news articles in both English and Persian are not appropriate, since usually English articles are very short and in most cases are translated summaries of Persian articles and besides their archives are not available online. On the other hand, articles from different news agencies are not easily aligned by date, since in many cases the same event is published in different dates. Thus we were made to choose two distinct collections in different origins and align the articles to obtain the comparable corpus using their publication dates and content similarity scores.

We follow the general procedure proposed in [4], [23] to construct our comparable corpus. Talvensaari et. el, in [23] present a method to build comparable corpora from two collections in different languages and different origins. In their work, they extract the best query keys from documents of one collection using the Relative Average Term Frequency (RATF) formula [17]. The keys are translated into the language of the other collection using a dictionary-based query translation program. The translated queries are run against the target collection and documents are aligned based on the similarity scores and their publication dates. Their method is tested on a Swedish newswire collection and an American newspaper collection. Their approach is most closely related to Braschler et. el, [4] which introduced a method to align documents in different languages by using dates, subject codes, common proper nouns, numbers and a small dictionary.

However, our method for constructing the corpus is different from [4], [23] in many details, somewhat because of the language differences and available linguistic resources. Talvensaari et. al [23], use TWOL lemmatizer [10] to lemmatize inflected Swedish document words and to decompose compound words, while we do not do any preprocessing on our Persian collection. Another difference is in translating the keywords. Talvensaari et. al used UTACLIR, a dictionary-based query translation program, which uses query structuring to disambiguate translation alternatives and a fuzzy string matching technique to transform words not found in the dictionary [23] to translate document keys, while we use a simple dictionary and Google translator[3] to translate missing words from dictionary.

---

[1] www.hamshahrionline.ir

[2] www.bbc.co.uk

[3] http://translate.google.com/

The existence of many non-translated English words in our work suggests that we should use transliteration. Using Google transliteration[4] is shown to be beneficial in our experiments. Moreover, Talvensaari et. al, use a lot of heuristics for their alignments and for setting the parameters and thresholds, while we try to be more general and do not include these heuristics in our work.

In our experiments, we evaluated several methods of creating comparable corpora and the experiment results show that (1) Using top-k translation alternatives of a word from dictionary can improve the quality of comparable corpus over using all translations of a word. (2) Using transliteration besides using dictionary and machine translation improves accuracy. (3) Using feedback retrieval model helps.

The rest of the paper is organized as follows. We explain the details of constructing our comparable corpus in section 2. We present the experiment results in section 3 and finally bring the conclusions and future work of our study in section 4.

2   Constructing the Comparable Corpus

To construct the comparable corpus, we start with two independent news collections, one in English and another in Persian, and try to align the documents in these collections. In our comparable corpus, we want the aligned documents to be as similar as possible. Intuitively, two documents are similar if their corresponding keywords - words that best describe the topic of the document - are close to each other. The publication date is another factor for finding good alignments. Documents with similar content which are published on the same date are most probably talking about the same event. Based on these intuitions, we follow these steps to align the documents: Extract the keywords of the documents in the source language, translate the keywords to the target language and run the translated queries against the target collection. We then align the documents based on their similarity scores and dates. In the rest of this section, we present the details of the method for constructing and evaluating the comparable corpus.

2.1   Query Construction and Translation

In order to construct query words for each source document, we applied the RATF formula [17].

$$RATF(k) = (cf_k/df_k) \times 10^3/ln(df_k + SP)^p, \tag{1}$$

In this formula, df_k and cf_k are document frequency and collection frequency of word k respectively. SP is a collection dependent scaling parameter to penalize rare words and p is a power parameter. We set these parameters to their best values reported in Talvensaari et. el, [23] which were SP = 1800 and p = 3. In

---

[4] www.google.com/transliterate/persian

order to construct queries which represent the source documents, we first sort the terms of each document in decreasing order of their frequencies within the document. Equal frequency keys are then sorted according to their RATF values. Finally, 30 top ranked keys are selected as the query which represents the document. Since not all the source language keys can be translated to the target language, we chose to select a slightly large number of keywords to represent each document. To translate selected keywords, we use an English-Persian Dictionary with more than 50,000 entries. Since there are lots of Out Of Vocabulary (OOV) words such as proper nouns, we use Google's machine translation system to translate the words not found in our dictionary. We then use Google transliteration system to translate the words which are still not translated, including some proper nouns and stemmed words.

After creating the query in the target language, we use a retrieval model to rank the target documents based on their similarities to the query. We use these similarity scores along with publication dates to align the documents.

2.2   Document Alignment

In order to create the alignment pairs, we use two basic criteria: the similarity scores of the documents and their publication dates. Intuitively, if an English document and a Persian document have a high similarity score and are published at the same date, they are likely talking about a related event. On the other hand, if the similarity score is low or the publication date is distant, the document pair doesn't seem to be a good match. We apply a combination of three different score thresholds $(\theta_1 < \theta_2 < \theta_3)$ to search for suitable document alignments. The values of thresholds are percentiles, e.g. $\theta = 60$ means that the score is greater than 60% of all the similarity scores in the runs. If there are n source documents and for each source document, r target documents are tested, then $n \times r$ scores should be considered to calculate the thresholds [23]. In our experiments, we set r to 50. The steps of document alignment are as follows. Considering an English document, we first search in its top r most similar target documents to find Persian documents which are published in the same day and also their similarity scores are greater that θ_1. If we couldn't find any alignment pair, the threshold increases to θ_2 and we search for target documents in a period of four days. Finally, if we still have not found any matching document, the score threshold is increased to θ_3 and a period of fourteen days is searched.

2.3   Comparable Corpora Evaluation

The main criterion for evaluating a comparable corpus is the quality of the alignments. In our experiments, we assessed the quality of alignments using a five-level relevance scale. The relevance scale is gained from [4]. The five levels of relevance are:

(1) Same story. Two documents are exactly about the same event.
(2) Related story. Two documents deal with same events but in somewhat different viewpoints. (e.g. one document may be part of the other document)

(3) Shared aspect. The documents cover two related events. (e.g. events in the same location or about same people)
(4) Common terminology. The similarity between the events is slight, but a considerable amount of terminology is shared.
(5) Unrelated. There is no apparent relation between the documents.

Which classes to be considered as good alignments depends on the intended application. For example, Braschler et. el, [4] considered classes (1) through (4) to be helpful for extracting good terms in CLIR systems. In our experiments, we count classes (1) through (3) as good alignments. Thus a high quality corpus is expected to have most of its alignments in levels (1), (2) and (3) and not many alignments in levels (4) and (5).

We also used other criteria for further evaluation of the corpus, for example the ability to extract meaningful word associations from the documents and the size of high quality discovered alignments.

In order to extract word associations from the comparable corpus, we used the method proposed in Talvensaari et. al, [24]. The intuition of this method is to use co-occurrence of words in the alignments to extract word associations. The algorithm first calculates a weight w_{ik} for each word s_i in document d_k as:

$$w_{ik} = \begin{cases} 0 & \text{if } tf_{ik} = 0 \\ (0.5 + 0.5 \times \frac{tf_{ik}}{Maxtf_k}) \times ln(\frac{NT}{dl_k}) & \text{otherwise} \end{cases} \tag{2}$$

where tf_{ik} is the frequency of s_i in document d_k , $Maxtf_k$ the largest term frequency in d_k and dl_k is the number of unique words in the document. NT can be either the number of unique words in the collection or its approximation. This $tf.idf$ modification is adopted from Sheridan and Ballerini [21] who also used it in similarity thesaurus calculation. The weight of a target word t_j in a set of ranked target documents D is calculated as:

$$W_j = \sum_{r=1}^{|D|} \frac{w_{jr}}{ln(r+1)}, \tag{3}$$

where D is the set of target documents aligned with a source document containing t_j. The documents in D are ranked based on their alignment scores. Less similar documents, which appear lower in the list, are trusted less for translation and their weights are penalized. This penalization is achieved by $ln(r+1)$ in the denominator.

Finally, the similarity score between a source word s_i and a target word t_j can be calculated as

$$sim(s_i, t_j) = \frac{\sum_{\langle d_k, D \rangle \in A} w_{ik} \times W_j}{\| s_i \| \times ((1 - \alpha) + \alpha \times \frac{\|t_j\|}{\|T\|})}, \tag{4}$$

in which w_{ik} is the weight of source word s_i in the source document d_k, W_j is the weight of target word t_j in the set of target documents D which are aligned with the source document d_k, A is the set of all alignments, $\| s_i \|$ is s_i's norm vector, $\overline{\| T \|}$ is the mean of the target term vector lengths, and α is a constant between 0 and 1 (we chose $\alpha = 0.2$ same as [24]). In this formula, Pivoted vector normalization scheme is employed to compensate long feature vectors.

3   Experiments and Results

In this section, we report our experiments on creating the Persian-English comparable corpora and the analysis of their qualities. In our experiments, we have used the Lemur toolkit[5] as our retrieval system. We used Porter stemmer for stemming the English words and Inquery's stopword list (418 words).

3.1   Document Collections

We have used news articles in Persian and English as our documents in Persian-English comparable corpora. Our English collection is composed of news articles published in BBC News and our Persian collection includes the news articles of Hamshahri newspaper. We have used five years of news articles, dated from Jan. 2002 to Dec. 2006. The BBC articles are crawled from the BBC News website and preprocessed to clean the web pages, and also to omit local news of United Kingdom, which will not be aligned with any Persian news article. The Hamshahri articles are extracted from Hamshahri corpus[6] which consists of all of the Hamshahri news articles published between 1996 and 2007. The details of the collections are given in Table 1.

Table 1. Statistics on the English and Persian Document Collections

Collection	# of Docs.	Time Span	Avg. Doc. Length	# Unique Terms
BBC	53697	Jan. 2002-Dec. 2006	461	141819
Hamshahri	191440	Jan. 2002-Dec. 2006	527	528864

3.2   Creating and Evaluating the Comparable Corpus

To construct the comparable corpus, we first extract the keywords of each document in the source language and translate the keywords to the target language (see section 2.1). These translations are considered as queries in the target language and are run against the target collection to retrieve a ranked list of related documents. The results are processed using the method explained in section 2.2 and tested with different document relevance score thresholds to create the document alignments and thus the comparable corpus.

We have experimented with different alternatives for (1) source language keyword translation, and (2) retrieval models. In order to compare the quality of

---

[5] http://www.lemurproject.org/
[6] http://ece.ut.ac.ir/dbrg/Hamshahri/

Table 2. Assessed Quality of Alignments in one month, Jan. 2002, of experiments with KL-divergence retrieval model. (a)using all dictionary translations of each word. (b)using top-3 translations of each word.

<table>
<tr><td colspan="3" align="center">(a)</td><td colspan="3" align="center">(b)</td></tr>
<tr><td></td><td># of Alignments</td><td>% of Alignments</td><td></td><td># of Alignments</td><td>% of Alignments</td></tr>
<tr><td>Class 1</td><td>4</td><td>11.76</td><td>Class 1</td><td>3</td><td>6.97</td></tr>
<tr><td>Class 2</td><td>4</td><td>11.76</td><td>Class 2</td><td>17</td><td>39.53</td></tr>
<tr><td>Class 3</td><td>7</td><td>20.58</td><td>Class 3</td><td>14</td><td>32.55</td></tr>
<tr><td>Class 4</td><td>11</td><td>32.35</td><td>Class 4</td><td>8</td><td>18.6</td></tr>
<tr><td>Class 5</td><td>8</td><td>23.52</td><td>Class 5</td><td>1</td><td>2.32</td></tr>
<tr><td>Total</td><td>34</td><td>100</td><td>Total</td><td>43</td><td>100</td></tr>
</table>

different alignments, corresponding to different comparable corpora, we manually assessed the quality of alignments for one month, Jan. 2002, on a five-level relevance scale (see section 2.3). Our evaluation results show that different alternatives for constructing the comparable corpora result in corpora with very different qualities.

In our first set of experiments, we used a simple dictionary to translate the source keywords and used Google translator to translate the query words not found in the dictionary. We used the KL-divergence retrieval model with pseudo relevance feedback as our retrieval model [11]. In our experiments, we set the score thresholds to $\theta_1 = 60, \theta_2 = 80$ and $\theta_3 = 95$. Table 2 (a) shows the assessed quality of alignments in this set of experiments using all available translations of a word in the dictionary. As the table shows, roughly 45% of the assessed alignments are about related events and more than half of the alignments are in classes (4) and (5) having little or no similarity.

Different words in our dictionary have different number of translations and this may bias our translated queries. In our second set of experiments, we just used the top three translations of each word in the dictionary to construct the query in the target language. Table 2 (b) shows the results of alignments in the second set of experiments. We can see that almost 79% of the alignments are about related events. This indicates, a big quality improvement by using top-3 translations of each word.

Using the dictionary accompanied with machine translator, there are still some source keywords which are not translated. These words are either proper nouns (such as Euroland, Brinkema, Moussaoui, Toiba and Belkheir) or stem words (such as Sydnei, Melbourn and Athen) which seem to have a high impact in our alignments. In our next set of experiments, we tried to transliterate those words which still are not translated. Table 3 shows the results of adding Google transliteration for missing words. We repeat the results with no transliteration for easier comparison. As can be seen from the table using transliteration can bring in better alignment pairs.

In our next set of experiments, we used Okapi with pseudo relevance feedback as our retrieval model [19]. We used the top 3 words of the dictionary for translation. Table 4 shows the results. Since our main goal is to find as many high

Table 3. Results of top-3 Translations With or Without Transliteration and KL-divergence Retrieval Model

	No Transliteration		Transliteration	
	# of alignments	% of alignments	# of alignments	% of alignments
Class 1	3	6.97	5	9.43
Class 2	17	39.53	24	45.28
Class 3	14	32.55	14	26.41
Class 4	8	18.6	8	15.09
Class 5	1	2.32	2	3.77
Total	43	100	53	100

Table 4. Results of top-3 Translation With or Without Transliteration and Okapi Retrieval Model

	No Transliteration		Transliteration	
	# of alignments	% of alignments	# of alignments	% of alignments
Class 1	11	13.58	13	14.94
Class 2	46	56.79	51	58.62
Class 3	20	24.69	19	21.83
Class 4	4	4.93	4	4.59
Class 5	0	0	0	0
Total	81	100	87	100

Table 5. Statistics on the Constructed Comparable Corpus

# of alignments	7580
# of unique English (BBC) documents	4267
# of unique Persian (Hamshahri) documents	3488
# of alignments in same day	1838
# of alignments in four day period	2786
# of alignments in fourteen day period	2956

quality document pairs as possible, we compare the size of the aligned corpora as well as their quality. As can be seen from tables 3 and 4, by using transliteration, the number of discovered alignments increases and interestingly, the newly discovered ones are mostly distributed in classes (1) and (2) which shows that most of the new aligned pairs are highly or fairly related.

This set of experiments has the best results among all our alignment experiments and we use this set of aligned documents as our comparable corpus. This comparable corpus consists of 7,580 document alignments. Using that specified thresholds, 8% of the 53,697 source documents are aligned. Table 5 shows some statistics about our created comparable corpus. Since the source and target documents are very different, the relatively low number of alignments was expected. Moreover, the number of alignments can be increased with lowering the thresholds, but this can also affect the quality of the comparable corpus. Table 6 shows

Table 6. Assessment of Alignment Quality for Two Different Sets of Score Thresholds

	$\theta_1 = 60, \theta_2 = 80, \theta_3 = 95$		$\theta_1 = 0, \theta_2 = 60, \theta_3 = 95$	
	# of alignments	% of alignments	# of alignments	% of alignments
Class 1	13	14.94	16	12.8
Class 2	51	58.62	62	49.6
Class 3	19	21.83	30	24
Class 4	4	4.59	14	11.2
Class 5	0	0	3	2.4
Total	87	100	125	100

Table 7. Top Term Similarities

English Word	Persian Word	Google Translation of Persian Word	Score	English Word	Persian Word	Google Translation of Persian Word	Score
iraqi	عراق	Iraq	104.91	korea	شمالی	North	78.86
korea	کره	Korea	93.73	market	بازار	Market	78.41
elect	انتخابات	Election	89.73	price	قیمت	Price	78.07
nuclear	هسته	Nucleus	85.02	economi	اقتصاد	Economy	77.77
champion	مدال	Medal	82.63	econom	اقتصادی	Economic	74.80
weapon	عراق	Iraq	82.15	oil	نفت	Oil	74.24
cancer	سرطان	Cancer	80.73	attack	حمله	Attack	73.02

the result with $\theta_1 = 0, \theta_2 = 60$ and $\theta_3 = 95$. As the table shows lowering the thresholds bring the 43.7% more aligned documents but percentage of high quality alignments drops from 95.4% to 86.4%. Our comparable corpus contains 76% of all the high quality alignments with very small number of low quality ones. The quality of alignments is crucial when extracting translation knowledge from the corpus.

3.3 Extracting Word Associations

As another criterion to examine the quality of our comparable collection, we tried to extract word associations from the corpus and assess the quality of obtained associations (see section 2.3). Naturally, the higher the quality of the comparable corpora, the more precise the word associations will be. Table 3.3 shows a sample set of top English-Persian associated word pairs extracted from our comparable corpus. We also include the Persian words' Google translations for the readers not familiar with Persian. As can be seen from the table, most of the matched words have a very high quality. We should note that we are showing the stemmed English words in this table and that's why some of the suffixes are missing.

In Table 8, we show the top Persian words aligned with four of the English words. As the table shows, the confidence of matching decreases as we go down the list but the word pairs are still related. This observation suggests that these results can be used in query expansion.

Table 8. Word Associations for Four English Words

English Word	Persian Word	Google Translation of Persian Word	Score	English Word	Persian Word	Google Translation of Persian Word	Score
korea	کره	Korea	93.73	cancer	سرطان	Cancer	80.73
	شمالی	North	78.86		بیماری	Disease	52.01
	پیونگ	Pyvng	71.77		بدن	Body	51.26
	ینگ	Yang	71.40		سلول	Cell	41.67
	جنوبی	South	61.35		مبتلا	Suffering	39.97
iraqi	عراق	Iraq	104.91	champion	مدال	Medal	82.63
	صدام	Saddam	95.05		المپیک	Olympics	82.20
	عراقی	Iraqi	82.97		قهرمان	Champion	73.50
	بغداد	Baghdad	82.78		قهرمانی	Championship	72.26
	حسین	Hussein	75.29		مسابقات	Competitions	72.17

3.4   Cross-Language Experiments

In the next step of our research, we intend to do cross-language information retrieval using the obtained cross-lingual word associations from the comparable corpus. As the cross-language information retrieval task we focus on the CLIR task of CLEF-2008[7]: Retrieval of Persian documents from topics in English. The document collection for this task contains 166,774 news stories (578MB) that appeared in the Hamshahri newspaper between 1996 and 2002. The queries are 50 topic descriptions in Persian and the English translations of these topics. The Persian queries are used for monolingual retrieval.

In this study, we use the top k associated words in Persian to translate a query word in English with the intuition that these translations are more reliable. We normalize the raw scores to construct translation probabilities and construct the corresponding Persian query language model for each English query. We then rank the documents based on the KL-divergence between the estimated query language models and the document language models. We use monolingual Persian retrieval as a baseline to which we compare the cross-language results. In our monolingual Persian runs, we only use the title field of each Persian query topic as the query words. Since there is not any reliable stemmer for Persian, we did not stem the Persian words. Table 9 shows the results of CC-Top-1, CC-Top-2 and CC-Top-5 translations, where we use the top 1, 2 and top 5 mined associated words from comparable corpora as the translations of each query word.

We also did another run of experiments for CLIR using a dictionary as our translation knowledge. We used the top 1, 3 and top 5 translations of each of the query words from the dictionary to translate the queries. As can be seen from the table, using only comparable corpora and compared to the monolingual baseline, we can achieve up to 33.3% of mean average precision, 39.35% of precision at 5 documents and 38.92% of precision at 10 documents. Using dictionary, we can achieve about 36.29% of mean average precision, 36.12% of precision at 5

---

[7] www.clef-campaign.org

Table 9. Query Translation using Comparable Corpus versus Dictionary

Method	MAP	% of Mono	Prec@5	% of Mono	Prec@10	% of Mono
Mono Baseline	0.42		0.62		0.596	
CC-Top-1	0.111	26.33	0.208	33.54	0.17	28.52
CC-Top-2	0.14	33.30	0.244	39.35	0.232	38.92
CC-Top-5	0.116	27.51	0.216	34.83	0.19	31.87
Dic-Top-1	0.12	28.46	0.212	34.19	0.18	30.20
Dic-Top-3	0.13	30.84	0.192	30.96	0.202	33.89
Dic-Top-5	0.153	36.29	0.224	36.12	0.206	34.56
Dic-all	0.139	32.97	0.2	32.25	0.184	30.87

and 34.56% of precision at 10 documents. These results show that using only comparable corpora as a translation resource to perform cross-language information retrieval is comparable to using dictionary naively, i.e., constructing the query in the target language by using all translations of each query word in the dictionary.

4   Conclusions and Future Work

In this work, we created a Persian-English comparable corpus, which is, to the best of our knowledge, the first big comparable corpus for Persian and English. We created the corpus from two independent news collections and aligned the documents based on their topic similarities and publication dates. We experimented with several alternatives for constructing the comparable corpora, such as different translation methods and different retrieval models. We assessed the quality of our corpus using different criteria. As the first and most important criterion, we used a five-level relevance scale to manually evaluate the quality of alignments for one month. The evaluation results show that by properly translating the query words and using Okapi with pseudo relevance feedback as the retrieval model, we can come up with a high quality comparable corpus, for which 95% of the assessed matched articles are highly or fairly about related events.

We also tried to extract word associations from the comparable corpus and evaluate the quality of obtained associations. Furthermore, we did cross-language information retrieval using the cross-lingual word associations extracted from the comparable corpus. Experiment results show promising results for extracting translation knowledge from the corpus, although it needs more exploration.

In our future work we are going to focus on CLIR task by improving the quality of extracted word associations. We will try to use the comparable corpus, along with other linguistic resources such as dictionaries, machine translation systems or parallel corpora to improve the CLIR performance. It will also be interesting to use the extracted translation knowledge to improve the quality of the created corpus, by using the extracted word associations as an additional resource to translate source language keywords.

Acknowledgments. This research is partially supported by Iran Telecommunication Research Center (ITRC).

References

1. AleAhmad, A., Amiri, H., Darrudi, E., Rahgozar, M., Oroumchian, F.: Hamshahri: A standard Persian text collection. Knowledge-Based Systems 22(5), 382–387 (2009)
2. Bekavac, B., Osenova, P., Simov, K., Tadić, M.: Making monolingual corpora comparable: a case study of Bulgarian and Croatian. In: LREC, pp. 1187–1190 (2004)
3. Bijankhan, M.: Role of language corpora in writing grammar: introducing a computer software. Iranian Journal of Linguistics (38), 38–67 (2004)
4. Braschler, M., Schäuble, P.: Multilingual information retrieval based on document alignment techniques. In: Nikolaou, C., Stephanidis, C. (eds.) ECDL 1998. LNCS, vol. 1513, pp. 183–197. Springer, Heidelberg (1998)
5. Collier, N., Kumano, A., Hirakawa, H.: An application of local relevance feedback for building comparable corpora from news article matching. NII. J. (Natl. Inst. Inform.) 5, 9–23 (2003)
6. Davis, M.W.: On the effective use of large parallel corpora in cross-language text retrieval. Cross-language Information Retrieval, 11–22 (1998)
7. Dimitrova, L., Ide, N., Petkevic, V., Erjavec, T., Kaalep, H.J., Tufis, D.: Multexteast: parallel and comparable corpora and lexicons for six central and eastern european languages. In: ACL, pp. 315–319 (1998)
8. Ghayoomi, M., Momtazi, S., Bijankhan, M.: A study of corpus development for Persian. International Journal of Asian Language Processing 20(1), 17–33 (2010)
9. Karimi, S.: Machine Transliteration of Proper Names between English and Persian. Ph.D. thesis, RMIT University, Melbourne, Victoria, Australia (2008)
10. Koskenniemi, K.: Two-level morphology: A general computational model for word-form recognition and production. Publications of the Department of General Linguistics, University of Helsinki 11 (1983)
11. Lafferty, J., Zhai, C.: Document language models, query models, and risk minimization for information retrieval. In: SIGIR, pp. 111–119 (2001)
12. McNamee, P., Mayfield, J.: Comparing cross-language query expansion techniques by degrading translation resources. In: SIGIR, pp. 159–166 (2002)
13. Miangah, T.M.: Constructing a Large-Scale English-Persian Parallel Corpus. Meta: Translators' Journal 54(1), 181–188 (2009)
14. Munteanu, D., Marcu, D.: Improving machine translation performance by exploiting non-parallel corpora. Comput. Linguist. 31(4), 477–504 (2005)
15. Oard, D., Diekema, A.: Cross-language information retrieval. Annual Review of Information Science and Technology 33, 223–256 (1998)
16. Pilevar, M.T., Feili, H.: PersianSMT: A first attempt to english-persian statistical machine translation. In: JADT (2010)
17. Pirkola, A., Leppanen, E., Järvelin, K.: The RATF formula (Kwok's formula): exploiting average term frequency in cross-language retrieval. Information Research 7(2) (2002)
18. Resnik, P.: Mining the web for bilingual text. In: ACL, pp. 527–534 (1999)
19. Robertson, S.E., Walker, S.: Some simple effective approximations to the 2-Poisson model for probabilistic weighted retrieval. In: SIGIR, pp. 232–241 (1994)

20. Sharoff, S.: Creating general-purpose corpora using automated search engine queries. In: WaCky! Working Papers on the Web as Corpus (2006)
21. Sheridan, P., Ballerini, J.P.: Experiments in multilingual information retrieval using the spider system. In: SIGIR, pp. 58–65 (1996)
22. Steinberger, R., Pouliquen, B., Ignat, C.: Navigating multilingual news collections using automatically extracted information. CIT 13(4), 257–264 (2005)
23. Talvensaari, T., Laurikkala, J., Järvelin, K., Juhola, M.: Creating and exploiting a comparable corpus in cross-language information retrieval. TOIS 25(4) (2007)
24. Talvensaari, T., Pirkola, A., Järvelin, K., Juhola, M., Laurikkala, J.: Focused web crawling in the acquisition of comparable corpora. Information Retrieval 11, 427–445 (2008)
25. Tao, T., Zhai, C.X.: Mining comparable bilingual text corpora for cross-language information integration. In: SIGKDD, pp. 691–696 (2005)
26. Utsuro, T., Horiuchi, T., Chiba, Y., Hamamoto, T.: Semi-automatic compilation of bilingual lexicon entries from cross-lingually relevant news articles on WWW news sites. In: Richardson, S.D. (ed.) AMTA 2002. LNCS (LNAI), vol. 2499, pp. 165–176. Springer, Heidelberg (2002)
27. Yang, C.C., Li, W., et al.: Building parallel corpora by automatic title alignment using length-based and text-based approaches. Information Processing & Management 40(6), 939–955 (2004)

Validating Query Simulators: An Experiment Using Commercial Searches and Purchases

Bouke Huurnink, Katja Hofmann, Maarten de Rijke, and Marc Bron

ISLA, University of Amsterdam, The Netherlands
{bhuurnink,k.hofmann,derijke,m.m.bron}@uva.nl

Abstract. We design and validate simulators for generating queries and relevance judgments for retrieval system evaluation. We develop a simulation framework that incorporates existing and new simulation strategies. To validate a simulator, we assess whether evaluation using its output data ranks retrieval systems in the same way as evaluation using real-world data. The real-world data is obtained using logged commercial searches and associated purchase decisions. While no simulator reproduces an ideal ranking, there is a large variation in simulator performance that allows us to distinguish those that are better suited to creating artificial testbeds for retrieval experiments. Incorporating knowledge about document structure in the query generation process helps create more realistic simulators.

1 Introduction

Search engine transaction logs offer an abundant source of information about search "in the wild." An increasing number of studies have been performed into how searches and result clicks recorded in transaction logs may be used to evaluate the performance of retrieval systems [6, 7, 9] . However, transaction logs often contain information that can be used to breach the privacy of search engine users. In addition, their content may contain commercially sensitive information. Therefore companies are reluctant to release such data for open benchmarking activities.

A solution may lie in using simulation to generate artificial user queries and judgments. Simulation is a method by which large numbers of queries and judgments may be obtained without user interaction, as a substitute for hand-crafted individual retrieval queries and explicitly judged sets of relevant documents. Simulated queries have been compared to manually created queries for information retrieval [2, 13]. Reproducing absolute evaluation scores through simulation has been found to be challenging, as absolute scores will change with, e.g., recall base. However, reproducing exact retrieval scores is not essential to developing a useful simulator when we wish to rank retrieval systems by their performance. Following this argument, the aim is to make a simulator that allows us to identify the *best performing* retrieval system.

Consequently, we assess simulation approaches based on how well they predict the relative performance of different retrieval systems. More precisely, we examine whether simulated queries and relevance judgments can be used to create an artificial evaluation testbed that reliably ranks retrieval systems according to their performance. To the best of our knowledge this question has not been addressed so far.

M. Agosti et al. (Eds.): CLEF 2010, LNCS 6360, pp. 40–51, 2010.

In the novel simulation approaches that we introduce, we integrate insights into users' search goals and strategies to improve the simulation, e.g., patterns regarding the items typically sought and/or purchased. We also enrich the simulation process by exploiting characteristics of the document collection in the commercial environment in which we work, in particular the fielded nature of the documents.

We assess the validity of the output of the simulation approaches that we consider by correlate system rankings produced by our simulator output data with a gold standard system ranking produced by real judgments. Ideally, a large number of judgments should be used to produce this gold standard ranking, so as to obtain stable average evaluation scores. As sets of manually judged queries are usually small and difficult to obtain, we use implicit judgments derived from the transaction log of a commercial environment, where users search for and purchase documents.

Purchase-query pairs (i.e., user queries and their subsequent purchases) can be identified from the transaction logs and we use these purchases as implicit positive relevance judgments for the associated queries [7]. In addition, we use a set of training purchase-query pairs to help inform our simulators. We use the simulators to create sets of artificial purchase-query pairs to use as evaluation testbeds. Each simulator is then assessed by determining whether its testbed produces similar retrieval system rankings to the gold standard testbed created using real-world (logged) purchase-query pairs.

The main contributions of this paper are:

1. A large-scale study of the correlation between system rankings derived using simulated purchase-query pairs and a system ranking obtained using real queries and (implicit) assessments.
2. Novel query simulation methods that exploit characteristics from real queries as well as document structure and collection structure.
3. A detailed analysis of factors that impact the performance of a simulator.

Our results can be used by creators of evaluation collections who have access to transaction log data but are unable to release it and by experimenters who want to create realistic queries for a document collection without having access to the transaction logs.

2   Related Work

We describe developments in creating simulated queries and relevance assessments for retrieval evaluation.

Research into simulation for information retrieval systems goes back to at least the 1980s. Early work focused on simulating not only queries and relevance assessments, but also the documents in a collection. Tague et al. [12] developed simulation models that produced output data similar to real-world data, as measured for example by term frequency distributions. However, retrieval experiments showed that evaluations using the simulated data did not score retrieval systems in the same way as evaluations using real-world data [13]. Gordon [4] suggested that simulators should focus on creating queries and relevance judgments for pre-existing (non-simulated) document collections: the remainder of the work that we describe has been performed under this condition.

A problem when simulating queries is to identify multiple relevant documents for a query. One solution has been to use document labels or pre-existing relevance judgments to identify sets of multiple related documents in a collection [1, 10, 11]. Queries are back-generated from the related documents, which are then considered relevant for that query. This allows queries to be defined so that they conform to predefined criteria, e.g., long vs. short. While queries generated in this manner can be used to evaluate retrieval systems, for validation purposes there is typically no user interaction data available. Therefore it is unclear how they compare to "real-world" queries.

An alternative to identifying multiple relevant documents per query is to focus on identifying a single relevant document per query. In the context of online search, Dang and Croft [3] work under this condition. They exploit hyperlinked anchor texts, using the anchor text as a simulated query and the hyperlink target as a relevant document. Azzopardi et al. [2] study the building of simulators for known-item queries—queries where the user is interested in finding a single specific document. Their approach is to first select a target known-item document from the collection, and to subsequently back-generate a query from the terms in the selected document. In both of these studies, simulator output was compared to real-world data with respect to validity for evaluating retrieval systems. Dang and Croft [3] modified the anchor text queries and compared system evaluations on the simulated modified queries to system evaluations on similarly modified queries and clicks taken from an actual search log. They found that, in terms of systems evaluation, the simulated queries reacted similarly to various modification strategies as real queries. Azzopardi et al. [2] compared their simulated known-item queries to sets of 100 manually created known-item queries, examining absolute evaluation scores attained by evaluating systems on the simulator output data. The simulator output sometimes produced retrieval scores statistically indistinguishable to those produced using real known-item queries. Factors found to affect the simulator assessments include the strategy used to select known items, and the strategy used to select query terms from a target document. The optimal simulator settings varied per document collection and per retrieval system.

Our work is similar to [2] as our goal is to assess simulators for retrieval evaluation. However, we focus on relative performance instead of absolute scores as we argue that this is a more feasible and useful goal. Instead of comparing simulators to explicit judgments for known-item queries, we compare our approaches to a large number of purchase-query pairs that are derived from implicit judgments obtained from a transaction log. We apply and extend simulation strategies developed in [2], and compare these to strategies that take characteristics of logged queries into account.

3   Simulation Framework

First, we describe our setup for assessing query/document pair simulators. Then we detail the simulators, including how they select target documents and query terms.

3.1   Validating Simulators

In validating simulators we are particularly interested in how closely the system rankings using simulated purchase-query pairs resemble the system rankings using real purchase-query pairs. To assess a simulator, we first run it to create an evaluation testbed consisting of a set of queries with associated relevant documents (one relevant document per query, thus resembling a known-item task). The relevance judgments thus generated are then used to score retrieval systems. We evaluate the performance of each retrieval system on the automatically generated testbed in terms of Mean Reciprocal Rank (MRR), a standard evaluation measure for known-item tasks. Next, systems are ranked by retrieval performance. The ranking is compared to the one obtained by ranking the systems using real purchase-query pairs from a commercial transaction log.

Comparisons between system rankings are couched in terms of the rank correlation coefficient, Kendall's τ. A "better" (read: more realistic) simulator would achieve a higher rank correlation score with the gold standard ranking. Kendall's τ has been used previously for determining the similarity of evaluation testbeds generated by different sets of human assessors. Voorhees [14] considered evaluation testbeds with $\tau \geq 0.9$ to be equivalent. An ideal simulator would produce exactly the same rankings as the real queries, i.e., $\tau = 1.0$.

3.2   Simulating Purchase Queries

We base our framework for simulating purchase-query pairs on the known-item search simulation framework presented in [2]. Purchase queries are similar to known-item searches in that, for both types of search, a single query is (usually) associated with a single relevant document. For purchase queries the user may not necessarily know beforehand exactly which item they wish to obtain, but usually purchases a single item. In our transaction log, 92% of the purchase queries led to exactly one purchased item.

We extend the framework in [2] by incorporating information about the document fields from which query terms are typically drawn (detailed below). We observe that users of search systems tend to select query terms from specific parts of a document, e.g., users in our setting have been found to frequently issue queries containing terms from document titles [8]. To take this into account, we allow the simulator to use information about document fields to select query terms.

The resulting simulator framework is summarized in Algorithm 1. First, we select a document d_p from the collection C that is considered to be the target document to be purchased, according to a distribution D_d. We determine the length of the query by drawing from a distribution D_l that is estimated using a random subset of the training data (a sample of real purchase queries). For each term to be added to the query, we then determine the field from which the term should be drawn according to distribution D_f, and finally sample a term from the associated term distribution D_t.

In this setting, the crucial problems become: (1) determining D_d, i.e., which document should to be the target of a query; (2) determining D_f, i.e., which field should be selected as the source for a query term; and (3) determining D_t, i.e., which terms should be used to search for the target document. We implement and compare multiple strategies for addressing each problem, which we will discuss in turn.

D_d: **Selecting target documents.**   We investigate the effect of varying the selection of documents to use as simulated purchases. In previous work selecting target documents according to document importance as captured by inlink counts was found to have a

---

Algorithm 1. Generalized overview of our purchase query simulator

---

Initialize an empty query $q = \{\}$
Select the document d_p to be the purchased document with probability $p(d_p)$
Select the query length l with probability $p(l)$
for i in $1 .. l$ **do**
    if Using field priors **then**
        Select the document field f from which a term should be sampled with probability $p(f)$
        Select a term t_i from the field model (θ_f) of f with probability $p(t_i|\theta_f)$
    else
        Select a term t_i from the document model (θ_d) with probability $p(t_i|\theta_d)$
    Add t_i to the query q
Record d_p and q to define the purchase-query pair

---

positive effect in obtaining scores closer to retrieval scores using "real" queries [2]. We operate in an environment where inlink information is not available. Therefore, we formulate two target selection strategies that are expected to be representative of a lower and upper bound in simulation accuracy: (1) a uniform selection strategy, and (2) an oracle selection strategy that selects documents that are known to have been purchased.

Uniform. All documents in the collection are equally likely to be selected (samples are drawn with replacement). This strategy only requires the presence of a document collection and does not assume additional information. The probability of selecting a document is $p_{\text{Uni}}(d_p) = |C|^{-1}$, where $|C|$ is the collection size.

Oracle. For each logged purchased document, a query is generated. This strategy exactly mirrors the distribution of purchased documents that is observed in the test collection. The probability of selecting a document is determined by: $p_{\text{Ora}}(d_p) = \#(d_p) \cdot (\sum_{d \in D_p} \#(d))^{-1}$, where $\#(\cdot)$ is the number of times a document is purchased and D_p is the set of purchased documents.

We expect the oracle selection strategy to result in a simulator for which the resulting system rankings more closely resemble a ranking resulting from real queries. If the two document selection strategies lead to large differences in correlations with a system ranking produced by real queries, this would mean that more complex strategies for generating purchase distributions should be investigated further.

D_t: Selecting query terms. The second step in developing a purchase query simulator is to generate query terms that a user might use to (re)find a given target document. Many strategies are possible — we focus on the effect of existing term selection methods and the effect of selecting terms from different document fields. The following selection methods, previously defined for known-item search in [2], are investigated:

Popular. Query terms are sampled from the purchased document using the maximum likelihood estimator. Frequently occurring terms in the document are most likely to be sampled. The probability of sampling a term is determined by: $p(t_i|\theta_d) = n(t_i, d) \cdot (\sum_{t_j \in d} n(t_j, d))^{-1}$, where $n(t, d)$ is the number of times t occurs in d.

Uniform. Query terms are sampled from the document using a uniform distribution (each term has an equally likely chance of being sampled): $p(t_i|\theta_d) = |d|^{-1}$, where $|d|$ is the number of unique terms in a document.

Discriminative. Query terms are sampled from the document using their inverse collection frequency. Terms that rarely occur in the collection are most likely to be sampled. The probability of sampling these terms is determined by:

$$p(t_i|\theta_d) = \frac{b(t_i, d)}{p(t_i) \cdot \sum_{t_j \in d} \frac{b(t_j,d)}{p(t_j)}}, \tag{1}$$

where $b(t, d) = 1$ if t is present in d and 0 otherwise and $p(t)$ is given by:

$$p(t_i) = \frac{\sum_{d \in C} n(t_i, d)}{\sum_{d \in C} \sum_{t_j \in d} n(t_j, d)}. \tag{2}$$

TF.IDF. Query terms are sampled from the document according to their TF.IDF value. Terms that occur rarely in the collection, but frequently in the document, are most likely to be sampled. Writing $df(t)$ for the document frequency of a term, we put:

$$p(t_i|\theta_d) = \frac{n(t_i, d) \cdot \log \frac{|C|}{df(t_i)}}{\sum_{t_j \in d} n(t_j, d) \log \frac{|C|}{df(t_j)}}. \tag{3}$$

Other term sampling strategies are possible, e.g., ones that take term distances or co-occurrences into account, but these go beyond the scope of this paper.

Sampling strategies are expected to affect how realistic simulated queries are, as they constitute different models of how users create query terms when thinking about a target document. Query formulation is more complex in real life, but a model that explains a large part of real query generation well will result in a better simulator.

A simulator that uses a term selection method that is close to that of the term scoring method underlying a retrieval system used to evaluate that simulator will score high on the queries thus generated. This need not result in a good simulator, as we are comparing system rankings to those resulting from evaluation using real purchase queries.

D_f: **Incorporating document structure.** Beyond the influence of the term selection strategy, we observe that users of online search systems tend to select query terms from specific parts of a document [8]. In the collection that we used for development (see Section 4), we observed the program description was the most frequent source of query terms, followed by the summary, title, and recording number fields. Table 1 gives a description of the fields that are used in our document collection, and specifies the likelihood that query terms are matched in the respective fields. We obtained the probabilities by matching terms in queries from the development set to their field locations in the associated purchased documents. If a term occurred in multiple fields, each field was counted once. Note that this analysis is not representative of the entire group of queries issued in the archive, as only certain types of queries have been included in this paper (see Section 4.)

In our simulation experiments, we systematically explore the use of document fields for generating simulated queries, using the four most frequent sources of query terms on the one hand, and the distribution of field priors on the other:

Table 1. Names and descriptions of fields available in our experimental document collection; $p(f)$ indicates the probability that the field contains a purchase query term

Name	Description	$p(f)$	Name	Description	$p(f)$
beschrijving	description of the program	0.4482	immix_docid	document id	0.0037
dragernummer	number of recording	0.1303	zendgemachtigde	broadcast rights	0.0020
hoofdtitel	program title	0.1691	rechten	copyright owner	0.0012
samenvatting	summary of the program	0.1406	genre	the type of a program	0.0008
selectietitel	titles of program sections	0.0348	dragertype	format of recording	0.0005
makers	creators of the program	0.0241	deelcatalogus	catalog sub-collection	0.0000
namen	names mentioned in the program	0.0190	dragerid	id of recording	0.0000
persoonsnamen	people in program	0.0153	dragersoort	type of recording	0.0000
geografische_namen	locations mentioned in program	0.0051	herkomst	origin of the program	0.0000
trefwoorden	keywords	0.0051	publid	publication id	0.0000

Whole Document. Query terms are sampled from the entire document without incorporating any information about fields.

Description. Query terms are sampled from the description field only.

Summary. Query terms are sampled from the summary field only.

Title. Query terms are sampled from the title field only.

Recording number. Query terms are sampled from the recording number field only.

Field priors. Query terms are drawn from any part of the document, but terms from fields that are more frequent sources of real query terms are more likely to be used as a source for query terms. This model corresponds to the full setup outlined in Algorithm 1, including the optional field selection step. Field prior probabilities are estimated using the development collection, and correspond to $p(f)$ in Table 1.

4 Experimental Setup

We describe the setup of our experiments in terms of the data used, the settings used when applying our simulators, and the retrieval systems used to assess the simulators.

4.1 Experimental Data

Our experimental data consists of a collection of documents and a large set of purchase-query pairs. We obtain this data from an audiovisual archive, the *Netherlands Institute for Sound and Vision*. The archive maintains a large collection of audiovisual broadcasts. Customers (primarily media professionals) can search for and purchase broadcasts (or parts of broadcasts). Their actions are recorded in transaction logs [8].

Each audiovisual broadcast in the archive is associated with a textual catalog entry that describes its content and technical properties. Search in the archive is based on text search of these entries. As our document collection we use a dump of the catalog entries made on February 1, 2010.

Our purchase-query pairs are derived from transaction log data gathered between November 18, 2008 and February 1, 2010. In some cases purchases are made for fragments of a broadcast; following the practice at TREC and other evaluation forums [5], we consider the entire broadcast relevant if it contains relevant information, i.e., if a fragment has been purchased. The transaction log includes queries that contain date filters and advanced search terms. We exclude these types of queries from our experiments, leaving their simulation for future work. In some cases a query resulted in purchases for multiple broadcasts, here we consider each purchase-query pair separately. In total we derived $31,237$ keyword-only purchase-query pairs from the collection.

Our documents and purchase-query pairs are divided into a training and a test set. The training set is used to derive probability estimates for simulation purposes. The test set is used to produce the gold standard ranking of retrieval systems. In order to preserve the same distribution of documents and purchase-query pairs in the training and test set, documents were randomly assigned to either set. Purchase-query pairs were then assigned to a set depending on the assignments of the purchased document.

4.2 Simulator Settings

We create a simulator for each combination of the target, term, and field selection models described in Section 3.2. Query lengths for the simulators are drawn from the distribution of query lengths in the training queries. Query terms are taken directly from documents without modification. We generate 1,000 purchase-query pairs per simulator. These are then used to evaluate the retrieval systems described below.

4.3 Retrieval Systems

To produce system rankings, we need multiple retrieval systems that generate diverse retrieval runs. To this end we use a variety of indexing and preprocessing methods and scoring functions. We use two open source toolkits for indexing and retrieval: Indri[1] (based on language modeling) and Lucene[2] (based on the vector-space model) to create a total of 36 retrieval systems. The main difference between the toolkits is the document scoring method, but they also differ in terms of pre-processing and indexing strategy; both are frequently used in real-world search applications and retrieval experiments. Some of the 36 retrieval systems are designed to give very similar performance, to capture subtle differences in ranking of similar systems; others are designed to give very different retrieval performance, to capture large fluctuations in performance.

We build three indexes, one using Indri, and two using Lucene. For the Indri index we use the standard pre-processing settings, without stemming or stop word removal. For the first Lucene index, we apply a standard tokenizer, for the second we additionally remove stop words and apply stemming using the Snowball Stemmer for Dutch.

The Indri retrieval toolkit is based on language modeling and allows for experimentation with different smoothing methods, which we use to generate runs based on a number of models. Documents receive a score that is based on the probability that the language model that generated the query also generated the document. This probability estimate is typically smoothed with term distributions obtained from the collection. The setting of these smoothing parameters can have a large impact on retrieval performance. Here we generate retrieval systems using Dirichlet smoothing with the parameter range $\mu = 50, 250, 500, 1250, 2500, 5000$. In this manner, we generate a total of

---

[1] http://www.lemurproject.org/indri/
[2] http://lucene.apache.org/

6 smoothing-based retrieval runs. Some of these 6 systems can be expected to produce similar retrieval results, allowing us to capture small differences in system rankings.

Both Indri and Lucene provide methods for indexing per field, allowing us to create alternative retrieval systems by forming different combinations of fields; Table 1 shows the names and descriptions of the indexed fields. We generate 10 fielded retrieval runs for each index (for a total of 30 runs), based on one or more of the following fields: *content* (all text associated with a document), *free(text)* (summary and description), *meta* (title and technical metadata), and *tags* (named entities, genre). The 10 field combinations can be expected to give very different performance, while applying a specific field combination to three index types will result in smaller variations in performance.

5   Results and Analysis

Our experiments are designed to validate simulation approaches by assessing how well their simulated purchase-query pairs rank retrieval systems in terms of their performance. A second goal is to identify the best performing simulator, i.e., the simulator that results in rankings that are closest to the gold standard ranking produced using real queries. In this section we provide an overview and analysis of our experimental results.

The correlation coefficients for the simulators produced in our experiments are given in Table 2. The simulator with the lowest coefficient of 0.286 uses *Discriminative* term selection in combination with *Summary* field selection and *Oracle* target selection, indicating that this simulator setting is particularly unsuited for generating realistic purchase-query pairs. The simulator with the highest correlation coefficient of 0.758 uses *Field Prior* field selection in combination with *Uniform* target selection, and *TF.IDF* term selection. None of the simulators achieves the value of $\tau \geq 0.9$ that indicates the equivalence of two testbeds created by human assessors, indicating that there is still plenty of room for improvement in creating simulators that realistically reproduce

Table 2. Correlation coefficients of system rankings using simulated queries and a system ranking using real-world data. The simulator with the highest coefficient overall is highlighted in bold. Shading has been used to differentiate between ranges of correlation coefficients: darkest shading for $\tau \geq 0.7$, medium shading for $0.5 \leq \tau < 0.7$, light shading for $0.3 \leq \tau < 0.5$, and no shading for $\tau < 0.3$.

	Uniform **Target Model**				*Oracle* **Target Model**			
	Term Model				*Term Model*			
Field Model	*Popular*	*Uniform*	*Discrim.*	*TF.IDF*	*Popular*	*Uniform*	*Discrim.*	*TF.IDF*
Whole Document	0.714	0.741	0.666	0.690	0.650	0.697	0.667	0.677
Description	0.393	0.348	0.347	0.360	0.382	0.373	0.371	0.375
Summary	0.352	0.340	0.365	0.355	0.435	0.476	0.286	0.384
Title	0.444	0.461	0.418	0.432	0.385	0.373	0.405	0.392
Recording Number	0.654	0.682	0.645	0.674	0.684	0.704	0.673	0.686
Field Priors	0.738	0.745	0.714	**0.758**	0.738	0.721	0.624	0.687

human querying and purchasing behavior. However, the large variance in simulator assessments does give some insight into which simulator strategies are preferable in the framework that we employ.

Incorporating field priors. Overall, we can see that simulators incorporating *Field Prior* field selection produce the most reliable system rankings, as measured by correlation to a system ranking using real purchase-query pairs. Except for one case, using field priors consistently and substantially improves over sampling from the whole document without taking field information into account.

The least reliable system rankings are produced by restricting field selection to a single field, as is the case for the *Title*, *Summary*, and *Description* field selection models. An exception is the *Recording Number* field. Simulators using this field selection model achieve correlations that are, depending on the term and target selection models, in many cases similar to the correlations achieved when using the whole document. This points to an important aspect of the real queries, namely, that many of them are high precision in nature. Professional users often know very well what they are looking for and how to find it, and searching on the highly distinctive recording number allows a document to be quickly targeted. Simulators missing this high-precision field lose out when trying to predict which retrieval systems will perform well in this setting.

Selecting known purchased documents. Simulators selecting documents that are known to be purchased (i.e., using *Oracle* target selection) generally do not produce better evaluation testbeds than simulators that select documents uniformly from the entire collection. This is somewhat surprising as Azzopardi et al. [2] found that a non-uniform sampling strategy based on document importance produced better simulators. However, the cited work validated simulators according to their ability to reproduce the absolute

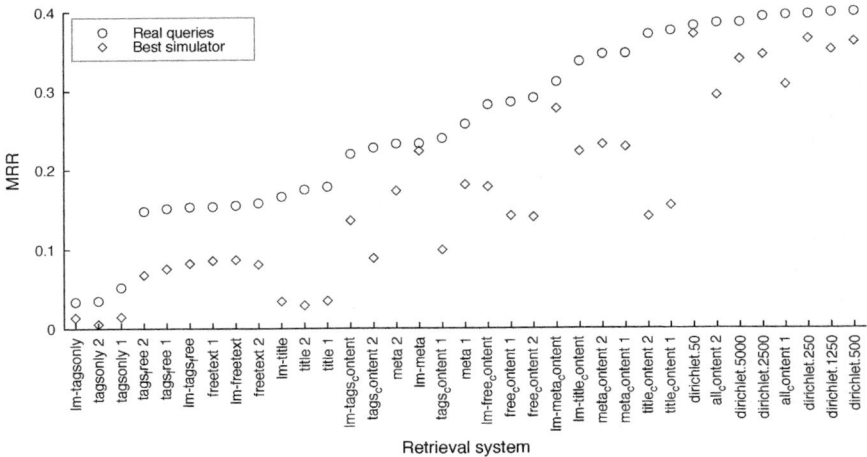

Fig. 1. Evaluation scores of retrieval systems in terms of MRR, on real purchase-query pairs derived from transaction logs, and on simulated purchase-query pairs generated by the best simulator ($D_d = Uniform$, $D_t = Uniform$, $D_f = FieldPriors$). Retrieval systems are sorted by their performance on real purchase-query pairs.

retrieval evaluation scores, while in this work we validate simulators according to their ability to rank retrieval systems. This may account for the difference in our findings, although further work is necessary to confirm this.

The impact of term selection. Comparing simulator assessment scores across the term selection models, we find subtle variations in correlation coefficients. No single model scores high consistently. In our experiments, then, the optimal term selection model is dependent on the target selection and field selection models that are used.

Absolute retrieval evaluation scores. For illustration purposes we briefly discuss absolute retrieval scores. Figure 1 plots the retrieval scores of retrieval systems evaluated with real purchase-query pairs from transaction logs, against the retrieval scores of the same systems when applied to the artificial purchase-query pairs generated by the best simulator. Some clusters of system rankings are correctly reproduced by the simulator (systems scoring better on real queries also tend to perform better on the simulated queries, and vice versa), even though absolute retrieval scores are generally lower than those obtained using the real queries.

6 Conclusion

We have explored the design and validation of simulators for generating purchase-query pairs, consisting of a query and an associated relevant purchased document, in a commercial setting. We developed a purchase query simulation framework incorporating new and previously existing simulation approaches. The framework incorporates models for (1) selecting target purchased documents, (2) selecting query terms, and (3) incorporating document structure in the query term selection process. By varying these models we created 48 simulators. Each simulator was used to produce an artificial evaluation testbed of simulated purchase-query pairs. The simulators were validated in terms of how well their testbeds ranked retrieval systems, as compared to the gold standard ranking obtained using a testbed of real logged purchase-query pairs.

No simulator produced an evaluation testbed that, according to the commonly accepted threshold for significance, ranked retrieval systems equivalently to the gold standard. This indicates that there is still plenty of room for improvement in the intrinsically difficult query simulation task. However, the large variation in simulator assessments did highlight broad categories of more successful and less successful approaches. Simulators were helped by explicitly including information about the parts of documents that query terms are drawn from in the term selection model. Further, in our setting, uniform selection of target purchased documents worked at least as well as a non-uniform selection of known purchased documents. This contrasts with previous findings in a different setting, where selecting "important" documents as targets for known-item searches resulted in improved simulator performance.

With large collections of queries being logged, one promising direction for improving simulators is further leveraging of the distributions of real-world queries and implicit judgments. For example, this study did not take the use of phrases as queries into account—their distribution could be estimated in sets of training queries and incorporated in a simulator. Other directions for future work include assessing the simulation approach in multilingual archival environments such as the European Library, and examining in further detail the multimedia aspect of this particular study, for example by exploring in detail the simulation of queries that target specific visual content.

Acknowledgements

This research was supported by: the European Union's ICT Policy Support Programme as part of the Competitiveness and Innovation Framework Programme, CIP ICT-PSP under grant agreement nr 250430; the DuOMAn project carried out within the STEVIN programme which is funded by the Dutch and Flemish Governments under project nr STE-09-12; the Center for Creation, Content and Technology (CCCT); the Dutch Ministry of Economic Affairs and Amsterdam Topstad under the Krant van Morgen project; and the Netherlands Organisation for Scientific Research (NWO) under project nrs 640.001.501, 612.066.512, 612.061.814, 612.061.815, 640.004.802.

References

[1] Azzopardi, L.: Query side evaluation: an empirical analysis of effectiveness and effort. In: SIGIR 2009, pp. 556–563. ACM, New York (2009)

[2] Azzopardi, L., de Rijke, M., Balog, K.: Building simulated queries for known-item topics: an analysis using six European languages. In: SIGIR 2007, pp. 455–462. ACM Press, New York (2007)

[3] Dang, V., Croft, B.W.: Query reformulation using anchor text. In: WSDM 2010, pp. 41–50. ACM Press, New York (2010)

[4] Gordon, M.D.: Evaluating the effectiveness of information retrieval systems using simulated queries. J. American Society for Information Science and Technology 41(5), 313–323 (1990)

[5] Harman, D.K.: The TREC test collection, chapter 2, pp. 21–52. TREC: Experiment and Evaluation in Information Retrieval (2005)

[6] He, J., Zhai, C., Li, X.: Evaluation of methods for relative comparison of retrieval systems based on clickthroughs. In: CIKM 2009, pp. 2029–2032. ACM, New York (2009)

[7] Hofmann, K., Huurnink, B., Bron, M., de Rijke, M.: Comparing click-through data to purchase decisions for retrieval evaluation. In: SIGIR 2010, Geneva, ACM, New York (July 2010)

[8] Huurnink, B., Hollink, L., van den Heuvel, W., de Rijke, M.: The search behavior of media professionals at an audiovisual archive: A transaction log analysis. J. American Society for Information Science and Technology 61(6), 1180–1197 (2010)

[9] Joachims, T.: Optimizing search engines using clickthrough data. In: KDD 2002, pp. 133–142. ACM, New York (2002)

[10] Jordan, C., Watters, C., Gao, Q.: Using controlled query generation to evaluate blind relevance feedback algorithms. In: JCDL 2006, New York, NY, USA, pp. 286–295. ACM, New York (2006)

[11] Keskustalo, H., Järvelin, K., Pirkola, A., Sharma, T., Lykke, M.: Test collection-based IR evaluation needs extension toward sessions–a case of extremely short queries. Inf. Retr. Technology, 63–74 (2009)

[12] Tague, J., Nelson, M., Wu, H.: Problems in the simulation of bibliographic retrieval systems. In: SIGIR 1980, Kent, UK, pp. 236–255. Butterworth & Co., Butterworths (1981)

[13] Tague, J.M., Nelson, M.J.: Simulation of user judgments in bibliographic retrieval systems. SIGIR Forum 16(1), 66–71 (1981)

[14] Voorhees, E.M.: Variations in relevance judgments and the measurement of retrieval effectiveness. In: SIGIR 1998, pp. 315–323. ACM Press, New York (1998)

Using Parallel Corpora for Multilingual (Multi-document) Summarisation Evaluation

Marco Turchi, Josef Steinberger, Mijail Kabadjov, and Ralf Steinberger

European Commission - Joint Research Centre (JRC), IPSC - GlobSec,
Via Fermi 2749, 21027 Ispra (VA), Italy
{Name.Surname}@jrc.ec.europa.eu
http://langtech.jrc.ec.europa.eu/

Abstract. We are presenting a method for the evaluation of multilingual multi-document summarisation that allows saving precious annotation time and that makes the evaluation results across languages directly comparable. The approach is based on the manual selection of the most important sentences in a cluster of documents from a sentence-aligned parallel corpus, and by projecting the sentence selection to various target languages. We also present two ways of exploiting inter-annotator agreement levels, apply them both to a baseline sentence extraction summariser in seven languages, and discuss the result differences between the two evaluation versions, as well as a preliminary analysis between languages. The same method can in principle be used to evaluate single-document summarisers or information extraction tools.

1   Introduction

In recent years, the large amount of daily contents produced in different languages has increased the need for more sophisticated Natural Language Processing (NLP) techniques that are able to summarise the main information across many languages. A clear example is the news domain, where hundreds of thousands of news articles are created every day, aggregated by news aggregators like Google News[1] and further analysed by systems like those of the Europe Media Monitor (EMM) family of applications [1], which extract and relate entities, link news clusters across languages and over time, and more.

Text mining technology – including entity recognition, sentiment analysis and summarisation – is available for English and a few other widely-spoken languages. The development of such tools for other languages is often hindered by the lack of training and evaluation resources, as creating these is labour-intensive and expensive.

Automatic summarisation is always evaluated by comparing automatic system output to a gold standard summary produced by human beings. This gold standard is created by reading and understanding documents and extrapolating a short and coherent abstract. As the selection of important facts or sentences is

---

[1] http://news.google.com

M. Agosti et al. (Eds.): CLEF 2010, LNCS 6360, pp. 52–63, 2010.
© Springer-Verlag Berlin Heidelberg 2010

a highly subjective process and a single summary cannot be considered statistically relevant, several independently produced summaries are typically used to evaluate any system output. Producing these is time-consuming and expensive.

Some human-annotated corpora are available for summarisation evaluation in English – the TAC data[2] is a valid example – but test data becomes the real bottleneck if the performance of an algorithm has to be computed in different languages. Even when such evaluation data exists for various languages, the test sets – and thus the evaluation results – are unlikely to be comparable across languages. To our knowledge, no parallel multilingual evaluation dataset exists for summarisation.

In this paper we focus on the specific challenges of testing multi-document summarisation algorithms in languages other than English, of making the results across languages comparable, and of saving development time, by making use of parallel documents.

Parallel corpora – texts and their exact translation – are widely used to train and evaluate Statistical Machine Translation systems. Some of the most widely known freely available parallel corpora are Europarl [2] and JRC-Acquis [3]. Many of them are from the domains of law and public administration.

Given a set of parallel and sentence-aligned documents in several languages referring to a particular topic (document cluster), our approach consists of manually selecting the most representative sentences in one of the languages (the pivot language). This sentence selection is then projected to all the other languages, by exploiting the parallelism property of the documents. The result is a multilingual set of sentences that can be directly used to evaluate extractive summarisation. When several annotators select sentences, the sentences can additionally be ranked, depending on the number of annotators that have chosen them.

Sentence selection is thus a crucial part of our methodology. A set of guidelines on how to select the sentences was given to each annotator. For each cluster, four annotators selected the most salient sentences, which were then scored according to the level of inter-annotator agreement. The score was applied to the output of an in-house summarisation algorithm, [4].

The created text collection cannot be distributed directly for copyright reasons. However, the meta-data is available for download at this url: `http://langtech.jrc.ec.europa.eu/JRC_Resources.html`.

2   Related Work

One of the bottlenecks of multilingual text mining tool development is the fact that any effort regarding evaluation procedures is typically multiplied by the number of languages covered. A common weakness of multilingual evaluation of text mining tools furthermore is that results are often not comparable across languages and domains.

---

[2] http://www.nist.gov/tac/2010/summarisation/index.html

In the effort to guarantee comparability of results across languages, various authors have suggested using multilingual parallel corpora: For instance, [5] proposed to exploit a multilingual bible corpus to get results that are more comparable for OCR than several multi-monolingual document sets. [6] used the multilingual parallel corpus EuroParl to evaluate Machine Translation performance across language pairs. The authors of [7] propose to use a multilingual parallel parsed corpus as the best and fairest gold standard for grammatical inference evaluation, because parallel documents can be assumed to have the same degree of language complexity.

[8] specifically worked on evaluation methods for multilingual multi-document text summarisation. They propose a whole range of evaluation methods, distinguishing human-human, human-computer and computer-computer agreement, using two classes of metrics: co-selection metrics (precision, recall, Kappa, and Relative Utility) and content-based metrics (simple cosine and cosine taking into consideration TF*IDF weighting of words; unigram- and bigram-overlap).

In the last decade, evaluation campaigns managed by DUC/TAC have taken a leading role in the summarisation roadmap. Research groups can compare their system performance with the state-of-the-art and obtain the expensive manual evaluation for free. For more details see http://www.nist.gov/tac/2010/summarisation/index.html.

The need of automatic evaluation metrics has increased in recent years. ROUGE [9], which measures n-gram co-occurrence between model summaries and the evaluated summary, has been used extensively. Although it correlates reasonably well with human assessments on average system scores, its average per-case correlation is low. Basic elements (BE) [10] use a similar idea, but on a syntactic level. An interesting method that still needs some human effort is the Pyramid approach [11]. However, all metrics have been applied to one language only, and no cross-language bootstrapping was performed. In comparison, we propose to use parallel corpora both to minimise the effort of evaluation resource creation and to ensure the comparability of results across languages. These ideas will be elaborated in the next sections.

3 An Approach to Multilingual Summarisation Evaluation

Our approach to multilingual extractive summarisation evaluation consists of two main ideas: (a) exploiting the nature of sentence-aligned parallel corpora so that a gold standard in one language can be projected to other languages, allowing thus to minimise the annotation effort; (b) using different degrees of inter-annotator agreement in the evaluation.

Projecting the summarisation gold-standard to other languages. In extractive summarisation, a single or multi-document summary is produced by selecting the most relevant sentences and by comparing these sentences with a gold standard of manually selected sentences. If sentence alignment information is available for a parallel text collection, the gold standard of one language can be projected to all the other languages. The more languages there are in the parallel corpus, the more time can be saved with this method.

Sentences are not always aligned one-to-one because a translator may decide, for stylistic or other reasons, to split a sentence into two or to combine two sentences into one. Translations and original texts are never perfect, so that it is also possible that the translator accidentally omits or adds some information, or even a whole sentence. For these reasons, aligners such as Vanilla[3], which implements the Gale and Church algorithm [12], typically also allow two-to-one, one-to-two, zero-to-one and one-to-zero sentence alignments. Alignments other than one-to-one thus present a challenge for the method of projecting the gold standard, in particular one-two and two-one alignments. We solved these issues in the following manner: When a manually selected gold standard sentence in language A is split into two sentences in language B, we add both sentences to the gold standard in language B. This choice increases the dimension of the gold standard, but it allows to maintain a certain coherence between languages A and B. In the opposite case, when the annotator selects a sentence in language A that is part of a pair of sentences that are the equivalent of a single sentence in language B, we add the human selected sentence to the gold standard in language A and the relative sentence in language B to the other gold standard. In our text collection, we found that less than 9% of the total number of alignments were not one-to-one. This general methodology could easily be applied in the evaluation of other NLP fields, in particular to entity extraction.

Using different degrees of inter-annotator agreement. Typically, two annotators do not produce the same gold standard annotation, especially for complex tasks such as the selection of the most important sentences. For this reason, it is useful to use more than one annotator. Various ways exist to deal with conflicting annotations: one can assume that only the intersection of all annotations is correct, or that each of the annotations is correct, or – as we do – one can weight the annotations depending on the number of annotators that have selected them.

In this work, we use four annotators. Each sentence is associated to a score ranging from zero to four, depending on the number of annotators that have selected that sentence. In our evaluation settings, the performance of the summarisation system is judged as better if the automatically selected sentences have also been manually selected by all or most of the annotators. A detailed description of the evaluation score is presented in Section 6.

Summary length unit: word or sentence? Most summarisation tasks (e.g. at TAC 2009) require the systems to produce summaries of a certain length (e.g. 100 words) rather than to select a certain number of sentences. This is reasonable because some sentences are simply much longer than others, and it is realistic because real-life systems are likely to be submitted to such a length restriction.

We decided to evaluate several numbers of selected sentences rather than summary lengths, for the following reasons: Our annotators were free to select

---

[3] http://nl.ijs.si/telri/Vanilla/

as many sentences as they deemed useful instead of creating a proper summary of a certain length. By giving the software the same task, the automatically and the manually produced results were equivalent and could be compared directly, without any loss of information. In our experiments, the summariser applied the content criterion, because the sentence length criterion potentially can lower the overall summary score. This means that a slightly longer summary may score lower compared to the maximally achievable summary score than a shorter summary not trying to reach the maximum summary length. However, if it were desirable for comparability reasons with other systems to produce summaries of a certain length, it is possible to first select sentences and to fill the remaining summary space with a relatively high-ranking summary sentence.

4 Parallel Document Evaluation Data

As described above, our approach takes advantage of the parallelism between documents in different languages. This is a big constraint that reduces a lot the available datasets.

In the next section, we describe the way we produce our collection of documents, and in particular the guidelines given to the human for the annotation phase.

4.1 The Texts

The documents of our dataset were downloaded from the Project Syndicate web site[4].

Creation of our dataset was carried out by the human annotators, who chose one subject each from the four main subject areas covered by Project Syndicate.[5] They came up with the following four subjects: *Israeli-Palestinian conflict, Malaria, Genetics* and *Science and Society*.

After that, each annotator was asked to select a homogeneous set of five related English language documents. Only texts existing in at least the seven languages English, French, Spanish, German, Arabic, Russian and Czech could be selected. The files were saved in UTF-8 format and the Corleone [13] sentence splitter was applied. The average number of sentences per document was over 50. In the "Malaria" cluster, there were roughly 54 sentences per document, in the "Israel" cluster 49 per document. This number of sentences per cluster is similar to that in the TAC setting.

The documents inside each cluster were aligned using Vanilla. Every non-English document was aligned with the English version of the same document, and all alignment information was stored in an xml file. We did not compute the

---

[4] http://www.project-syndicate.org/. Project Syndicate is a voluntary, member-based institution that produces high quality commentaries and analyses of important world events. Each contributor produces a commentary in one language. This is then human-translated into various other languages.

[5] That is, from the areas listed on the left-hand side of the Project Syndicate website.

alignment between non-English documents. In total, there were 91.7% of one-to-one alignments, 0.3% of type zero-to-one, 3.4% two-to-one, 4.49% one-to-two and 0.2% two-to-two.

As a result, we had four clusters of parallel documents in seven languages, plus the corresponding alignment information.

4.2 Human Annotation

Four human subjects were chosen for the annotation, three with a Computer Science background and one with a Linguistics background. After the selection of the clusters and documents, all four annotators were asked to read and label, independently of each other, all "summary-worthy" sentences from the cluster on "Israeli-Palestinian conflict". To allow for maximum freedom of choice, the concept of "summary-worthy" sentences was purposefully loosely defined as being prominent sentences.

After this pilot annotation phase preliminary inter-annotator agreement was computed and the annotators were asked to meet and discuss their experience and insights. The outcome of the annotation was the following: 10 sentences were chosen by all four annotators, 11 by at least three, 27 by at least two, 42 by at least one and 102 were not selected, see Table 1.

Following joint discussions among the annotators, the definition of "summary-worthy" sentence was refined as follows. When labelling a sentences as "summary-worthy":

- Keep in mind both cluster-level and title relevance/relatedness/salience.
- If a sentence is chosen:
  - double-check that it is NOT an empty sentence in itself (i.e., it conveys sufficient information of the reason why you are selecting it), otherwise, include only minimal and essential context (e.g., for key pronouns include closest sentence with antecedent).
- Regarding content analysis:
  - look for essential background;
  - endeavour to preserve the authors' point of view;
  - try preserving argument structure.

With these guidelines in mind, the four human subjects were asked to annotate a second cluster, the one on "Malaria". Four annotators guarantee reliability of

Table 1. Relative agreement among all 4 annotators (Number of sentences and percentage)

	Israel	Malaria	Average
Selection agreement of all annotators	10 (5%)	6 (3%)	8 (4%)
Selection agreement of 3 annotators	11 (6%)	10 (5%)	10.5 (5%)
Selection agreement of 2 annotators	27 (14%)	21 (10%)	24 (12%)
Selection by only 1 annotator	42 (22%)	51 (23%)	46.5 (23%)
Non-selection agreement of all annotators	102 (53%)	129 (59%)	115.5 (56%)

Table 2. Average agreement of any pair of the four annotators(Number of sentences and percentage)

	Israel	Malaria	Average
Selection agreement	20 (10%)	14.5 (7%)	17.25 (8%)
Selection of sentences by 1 annotator	44.5 (23%)	44.5 (21%)	44.5 (22%)
Non-selection agreement	127.5 (67%)	158 (72%)	142.75 (70%)

the selection approach across cluster domains. The inter-annotator agreement is discussed in the following subsection.

4.3   Inter-annotator Agreement

In General, summary production is a very subjective task. We looked at the agreement from two different points of view: the relative agreement of all four annotators (Table 1), and the average agreement of any pair of the four annotators (Table 2).

The number of sentences that were selected on average by a pair of annotators (first row of Table 2) is obviously bigger than the intersection of sentences selected by all the four annotators (first row of Table 1). On the other hand, the total number of sentences selected by at least two annotators (first three rows of Table 1) creates a larger basket of sentences for the evaluation score rather than the first row of Table 2. In total, annotators agreed to include 21% of sentences (first three rows of Table 1) compared to only 8% on those of the pairs of annotators. It means that the first point of view produces a steeper pyramid of agreements with a smaller top, but fine-grained discrimination capability due to the higher number of levels. On the other hand, the second one has a moderate pyramid of agreements with a larger top, but a coarse discrimination capability.

In both the tables, we can see that the "Israel" cluster seems to be more compact than the "Malaria" one. In fact, the annotators had an agreement on a large number of sentences. This result shows how difficult the selection of coherent documents in the same cluster is.

4.4   Copyright Issue and Distribution of the Evaluation Data

Unfortunately, for property right reasons, we cannot make the test collection directly available to peers. We did not get an answer to our requests to receive the right to distribute the evaluation data for scientific research purposes. For this reason, we decided to distribute only the metadata for the documents and to leave it to each interested party to download and use the documents. The metadata consists of a list of URLs where the documents can be found, a list of sentence offsets (sentence borders), the manual sentence selection of all annotators, as well as the sentence alignment information across languages. The meta-data is available for download.

5   Multilingual Multi-document Summarisation

The focus of this paper is to present the evaluation methodology. Therefore, our baseline sentence extraction summarisation system will only be presented briefly.

Originally proposed by Gong and Liu [14] and later improved by J. Steinberger and Jezek, [15,4], this approach first builds a term-by-sentence matrix from the source, then applies Singular Value Decomposition (SVD) and finally uses the resulting matrices to identify and extract the most salient sentences. SVD finds the latent (orthogonal) dimensions, which in simple terms correspond to the different topics discussed in the documents. As features it used in addition to unigrams also bigrams that occurred at least twice in the source. This approach was evaluated among the top ones at TAC'09[6].

The only language-dependent part of the summariser is the removal of terms in the stopword list. For each language we used different lists of stopwords with different length: Arabic 1092 words, Czech 804, German 1768, English 579, Spanish 743, French 1184, Russian 1004. We did not use any stemming or lemmatization.

6   Results and Discussion

In this section we are going to discuss the methodology for comparing system summaries against the model summaries produced by annotators. We report results of summaries with three different lengths - 5, 10, and 15 sentences on all 7 languages. We compared our LSA-based summariser with a Random summariser[7] and a Lead summariser, which selects the first sentences from each article. In each cluster there were 5 articles, thus the Lead summaries contain the first sentence of each article when the summary length is 5 sentences, the first two sentences of each article in the case of 10-sentence summaries, and the first three for 15-sentence summaries.

Weighted Model.   Firstly, we took the agreement of all annotators and we created a weighted model: weight 4 was assigned to sentences that were selected by all annotators, weight 3 to sentences selected by 3 annotators, and similarly weight 2 and 1. Then, for each summary, we computed the intersection of sentences selected by the summariser with those in the weighted model. For each sentence in the automatically generated summary, we added the model summary weight to the summary score. This makes the computation similar to Relative Utility discussed in [8]. We normalized this score by the maximal score that could be reached. The ratio of the summary score and the maximal possible score gave the final percentage similarity score reported in Tables 3-5. There we can see the performance of the summarisers on three different summary lengths in all analyzed languages together with the average performance.

---

[6] At TAC'09, the approach used more types of features. In addition to unigrams and bigrams it used person/organization/location entities and MeSH terms [16].

[7] We computed the results of the Random summariser based on Multivariate hypergeometric probability distribution.

Table 3. Summary evaluation using weighted model - summary length is 5 sentences

	Rnd	Lead	LSA
ar	20%	33%	40%
cz	19%	33%	45%
de	20%	33%	40%
en	19%	33%	38%
es	19%	33%	33%
fr	20%	33%	45%
ru	21%	33%	45%
AVG	20%	33%	41%

Table 4. Summary evaluation using weighted model - summary length is 10 sentences

	Rnd	Lead	LSA
ar	21%	27%	44%
cz	20%	26%	39%
de	21%	21%	40%
en	20%	28%	42%
es	20%	30%	41%
fr	21%	26%	41%
ru	21%	27%	36%
AVG	21%	27%	40%

Table 5. Summary evaluation using weighted model - summary length is 15 sentences

	Rnd	Lead	LSA
ar	23%	28%	45%
cz	22%	28%	43%
de	23%	26%	37%
en	22%	28%	39%
es	22%	27%	35%
fr	22%	28%	42%
ru	23%	28%	45%
AVG	22%	28%	42%

We can observe that the LSA summariser performs better than the Lead, which is better than Random[8]. We do not discuss here if the differences are statistically significant as we analyzed only two clusters. The performance differs from language to language. However there are not any significant peaks.

If two summarisers select two sets of sentences: the first set contains a sentence selected by all the annotators and one that is not selected at all, the general score will be $\frac{4+0}{4+4} = 0.5$. The second contains two sentences that are annotated by only two annotators, in this case the score will be $\frac{2+2}{4+4} = 0.5$. This example raises two questions: Would a human being prefer to get a two-two summary (second case in the example) rather than a four-zero summary (first case in the example)? and are the sentences at the top level twice more important than those selected by two annotators, or vice-versa?

Binary Model. To investigate these open questions, we used a more compact sentence scoring approach: a sentence was found important if it was selected by *at least* two annotators (binary model). The number of annotators can vary according to the number of the available annotators and the expected summary length. For the proposed dataset, we considered that two annotators were a reasonable choice and that the agreement of two of them on a single sentence was a reliable indication of importance.

This results in a binary quality level, a sentence can only be important or unimportant. The score of a system summary can then be computed as: the number of sentences in the intersection between the system summary and the sentences selected by at least two annotators divided by the number of sentences in the system summary. Tables 6-8 show the results of this evaluation approach.

We can see, that for example in the case of 5-sentence summaries, the LSA summariser selected on average 3 sentences (60%) that at least two annotators marked as important. We can also observe that the performance of Lead went down and, on the contrary, the performance of LSA went up in comparison with

---

[8] We did not do any tuning of the summariser on the analyzed type of texts. We used the TAC'09 settings of the summariser.

Table 6. Summary evaluation using binary model - summary length is 5 sentences

	Rnd	Lead	LSA
ar	22%	30%	50%
cz	21%	30%	70%
de	22%	30%	70%
en	21%	30%	60%
es	21%	30%	50%
fr	21%	30%	60%
ru	24%	30%	60%
AVG	22%	30%	60%

Table 7. Summary evaluation using binary model - summary length is 10 sentences

	Rnd	Lead	LSA
ar	22%	25%	60%
cz	21%	25%	70%
de	22%	20%	55%
en	21%	25%	60%
es	21%	30%	50%
fr	21%	25%	45%
ru	24%	25%	50%
AVG	22%	25%	56%

Table 8. Summary evaluation using binary model - summary length is 15 sentences

	Rnd	Lead	LSA
ar	22%	27%	53%
cz	21%	27%	53%
de	22%	23%	43%
en	21%	27%	47%
es	21%	27%	37%
fr	21%	27%	47%
ru	24%	27%	57%
AVG	22%	26%	48%

the previous approach. The difference in performance are given by the weighting model, because the summariser produces the same summary. In fact, not all the sentences at the beginning of the articles are selected by all the annotators, so the Lead performance decreases. On the contrary, the summariser that selects more often sentences chosen by more annotators rather than unimportant sentences, gets a higher score.

In particular, in the example proposed above, the first set gets a score equal to 0.5 $(\frac{1+0}{1+1} = 0.5)$, while the second one equals to 1 $(\frac{1+1}{1+1} = 1)$. It is unquestionable that the choice of the best set is arbitrary, but at least the binary model is able to disambiguate it in favour of the two-two selection. This led us to consider that the binary weighting model has a better discriminative power.

However, the binary model raises the question whether there is any need to use four annotators, whether two annotators would probably be enough. To get closer to the answer we computed another set of results, using a different binary diversification (We did not add the table due to lack of space). For each pair of annotators, we accepted as gold standard those sentences that were selected by both annotators. We used this set of sentences for the evaluation of the automatic summariser, and we then computed the average results for all six possible annotator pairs. The results show that relations between the three summarisers, Random < Lead < LSA, are maintained, although the differences between them decrease. Furthermore, the number of sentences in the gold standard selection is smaller than the number of sentences in the four-annotator schema (see Section 4.3 and tables 1 and 2). This suggests that the binary model with two annotators has less discriminative power – few summaries will have high scores and a lot of summaries bad scores – and less resistance to random hits. We can conclude that a higher number of annotators is always preferable to a smaller one.

Comparison across languages. We are also interested in understanding how the summariser performs across different languages, and whether it selects the same sentences. We found that the summariser selects on average only 35% of the same sentences for a language pair. Agreement peaks do exist, like the

Czech-Russian pair (41%), which may be due to the fact that they are both Slavic languages and thus have similar properties. The language pairs with least sentence selection agreement were Arabic-German (28%), Russian-German (27%), Russian-French (27%), and Russian-Spanish (28%). The fact that, overall, the sentence selection agreement across languages is quite so low indicates that there is a real need for multilingual summarisation evaluation, even if the summariser in principle uses only statistical, language-independent features. Note that this kind of analysis was not possible before due to the lack of multilingual parallel evaluation data.

7   Conclusion and Future Work

In this work, we addressed the problem of the evaluation of automatically generated summaries in languages other than English and the lack of freely available datasets for this purpose.

We proposed a semi-automatic approach to generate corpora for research on multilingual summarisation taking advantage of the parallelism among documents in different languages. In our methodology, all the gold standards in various languages were generated projecting a set of human-selected English sentences to other languages. The more languages there are in the parallel corpus, the more time can be saved with this method.

We proposed also an evaluation score based on the different degrees of inter-annotator agreement between human annotators. Two variations of the score have been investigated: the weighted and binary models, showing the more discriminative power of the binary model. For the first time, to our knowledge, the performance of automatic summarisers on seven different languages were compared.

Due to the nature of parallel corpora, this evaluation method can in principle also be applied to evaluate other text mining tools such as information extraction systems, but this has not yet been tested. In particular it can be used for entity extraction evaluation, as we can assume that each entity mention is present in each of the translations, meaning that each entity in the gold standard language will be present in all the other languages.

The produced data is available for download in a meta-data format. The full text collection cannot be distributed due to copyright issues.

References

1. Steinberger, R., Pouliquen, B., van der Goot, E.: An Introduction to the Europe Media Monitor Family of Applications. In: Information Access in a Multilingual World workshop at SIGIR, Boston, USA, pp. 1–8 (2009)
2. Koehn, P.: Europarl: A Parallel Corpus for Statistical Machine Translation. In: X Machine Translation Summit, Phuket, Thailand, pp. 79–86 (2005)
3. Steinberger, R., Pouliquen, B., Widiger, A., Ignat, C., Erjavec, T., Tufis, D., Varga, D.: The JRC-Acquis: A multilingual aligned parallel corpus with 20+ languages. In: LREC, Genova, Italy, pp. 24–26 (2006)

4. Steinberger, J., Ježek, K.: Update summarisation based on Latent Semantic Analysis. In: TSD, Pilsen, Czech Republic (2009)
5. Kanungo, T., Resnik, P.: The Bible, truth, and multilingual OCR evaluation. International Society for Optical Engineering, 86–96 (1999)
6. Koehn, P.: Europarl: A Multilingual Corpus for Evaluation of Machine Translation, unpublished draft (2002)
7. Van Zaanen, M., Roberts, A., Atwell, E.: A multilingual parallel parsed corpus as gold standard for grammatical inference evaluation. In: The Amazing Utility of Parallel and Comparable Corpora Workshop, pp. 58–61 (2004)
8. Radev, D., Allison, T., Blair-Goldensohn, S., Blitzer, J., Celebi, A., et al.: MEAD-a platform for multidocument multilingual text summarisation. In: LREC, Lisbon, Portugal, pp. 86–96 (2004)
9. Lin, C., Hovy, E.: Automatic evaluation of summaries using n-gram co-occurrence statistics. In: HLT-NAACL, Edmonton, Canada, pp. 71–78 (2003)
10. Hovy, E., Lin, C., Zhou, L.: Evaluating duc 2005 using basic elements. In: DUC 2005 (2005)
11. Nenkova, A., Passonneau, R.: Evaluating content selection in summarisation: The pyramid method. In: NAACL, Boston, USA (2004)
12. Gale, W.A., Church, K.W.: A program for aligning sentences in bilingual corpora. Computational Linguistics 19, 75–102 (1994)
13. Piskorski, J.: CORLEONE-Core Linguistic Entity Online Extraction. Technical report EUR 23393 EN, European Commission (2008)
14. Gong, Y., Liu, X.: Generic text summarisation using relevance measure and latent semantic analysis. In: ACM SIGIR, New Orleans, US, pp. 19–25
15. Steinberger, J., Ježek, K.: Text summarisation and singular value decomposition. In: Yakhno, T. (ed.) ADVIS 2004. LNCS, vol. 3261, pp. 245–254. Springer, Heidelberg (2004)
16. Steinberger, J., Kabadjov, M., Pouliquen, B., Steinberger, R., Poesio, M.: WB-JRC-UT's Participation in TAC 2009: Update summarisation and AESOP Tasks. In: TAC, NIST (2009)

MapReduce for Information Retrieval Evaluation: "Let's Quickly Test This on 12 TB of Data"

Djoerd Hiemstra and Claudia Hauff

University of Twente, The Netherlands

Abstract. We propose to use MapReduce to quickly test new retrieval approaches on a cluster of machines by sequentially scanning all documents. We present a small case study in which we use a cluster of 15 low cost machines to search a web crawl of 0.5 billion pages showing that sequential scanning is a viable approach to running large-scale information retrieval experiments with little effort. The code is available to other researchers at: `http://mirex.sourceforge.net`

1 Introduction

A lot of research in the field of information retrieval aims at improving the *quality* of search results. Search quality might for instance be improved by new scoring functions, new indexing approaches, new query (re-)formulation approaches, etc. To make a scientific judgment of the quality of a new search approach, it is good practice to use benchmark test collections, such as those provided by CLEF [10] and TREC [13]. The following steps typically need to be taken: *1)* The researcher codes the new approach by adapting an existing search system, such as Lemur [6], Lucene [8], or Terrier [11]; *2)* The researcher uses the system to create an inverted index on the documents from the test collection; *3)* The researcher puts the queries to the experimental search engine and gathers the top X search results (a common value for CLEF experiments is $X = 1000$); *4)* The researcher compares the top X to a golden standard by computing standard evaluation measures such as mean average precision. In our experience, Step 1, actually coding the new approach, takes by far the most effort and time when conducting an information retrieval experiment. Coding new retrieval approaches into existing search engines like Lemur, Lucene and Terrier is a tedious job, even if the code is maintained by members of the same research team. It requires detailed knowledge of the existing code of the search engine, or at least, knowledge of the part of the code that needs to be adapted. Radical new approaches to information retrieval, i.e., approaches that need information that is not available from the search engines inverted index, require reimplementing part of the indexing functionality. Such radical new approaches are therefore not often evaluated, and most research is done by small changes to the system.

Of in total 36 CLEF 2009 working note papers that were submitted for the Multilingual Document Retrieval (Ad Hoc) and Intellectual Property (IP) tracks,

M. Agosti et al. (Eds.): CLEF 2010, LNCS 6360, pp. 64–69, 2010.
© Springer-Verlag Berlin Heidelberg 2010

10 papers used Lemur, 10 used Lucene, 3 used Terrier, 1 used all three, 5 did not mention the engine, and 7 used another engine (Cheshire II, Haircut, Open Text SearchServer, MG4J, Zettair, JIRS, and MonetDB/HySpirit). In several cases a standard engine was used, and then a lot of post-processing was done on top, completely changing the retrieval approach in the end. Post-processing on top of an existing engine tests the ability of new retrieval approaches to rerank results from the existing engine, but it does not test the ability of new approaches to find *new* information.

In his WSDM keynote lecture, Dean [3] describes how MapReduce [4] is used at Google for experimental evaluations. New ranking ideas are tested off-line on human rated query sets similar to the queries from CLEF and TREC. Running such off-line tests has to be easy for the researchers at Google, possibly at the expense of the efficiency of the prototype. So, it is okay if it takes hours to run for instance 10,000 queries, as long as the experimental infrastructure allows for fast and easy coding of new approaches. A similar experimental setup was followed by Microsoft at TREC 2009: Craswell et al. [2] use DryadLINQ [15] on a cluster of 240 machines to run web search experiments. Their setup sequentially scans all document representations, providing a flexible environment for a wide range of experiments. The researchers plan to do many more to discover its benefits and limitations.

The work at Google and Microsoft shows that sequential scanning over large document collections is a viable approach to experimental information retrieval. Some of the advantages are:

1. Researchers spend less time on coding and debugging new experimental retrieval approaches;
2. It is easy to include new information in the ranking algorithm, even if that information would not normally be included in the search engine's inverted index;
3. Researchers are able to oversee all or most of the code used in the experiment;
4. Large-scale experiments can be done in reasonable time.

We show that indeed sequential scanning is a viable experimental tool, even if only a few machines are available. In Section 2 we describe the MapReduce search system. Sections 3 and 4 contain experimental results and concluding remarks.

2   Sequential Search in MapReduce

MapReduce is a framework for batch processing of large data sets on clusters of commodity machines [4]. Users of the framework specify a *mapper* function that processes a key/value pair to generate a set of intermediate key/value pairs, and a *reducer* function that processes intermediate values associated with the same intermediate key. The pseudo code in Figure 1 outlines our sequential search implementation. The implementation does a single scan of the documents, processing all queries in parallel.

```
mapper (DocId, DocText) =
  FOREACH (QueryID, QueryText) IN Queries
    Score = experimental_score(QueryText, DocText)
    IF (Score > 0)
    THEN OUTPUT(QueryId, (DocId, Score))

reducer (QueryId, DocIdScorePairs) =
  RankedList = ARRAY[1000]
  FOREACH (DocId, Score) IN DocIdScorePairs
    IF (NOT filled(RankedList) OR
      Score > smallest_score(RankedList))
    THEN ranked_insert(RankedList, (DocId, Score))
  FOREACH (DocId, Score) IN RankedList
    OUTPUT(QueryId, DocId, Score)
```

Fig. 1. Pseudo code for linear search

The *mapper* function takes pairs of *document identifier* and *document text* (DocId, DocText). For each pair, it runs all benchmark queries and outputs for each matching query the *query identifier* as key, and the pair *document identifier* and *score* as value. In the code, Queries is a global constant per experiment. The MapReduce framework runs the mappers in parallel on each machine in the cluster. When the map step finishes, the framework groups the intermediate output per key, i.e., per QueryId. The *reducer* function then simply takes the top 1000 results for each query identifier, and outputs those as the final result. The reducer function is also applied locally on each machine (that is, the reducer is also used as a *combiner* [4]), making sure that at most 1000 results have to be sent between machines after the map phase finishes.

3   Case Study: ClueWeb09

The ClueWeb09 test collection consists of 1 billion web pages in ten languages, collected in January and February 2009. The dataset is used by several tracks of the TREC conference [13]. We used the English pages from the collection, about 0.5 billion pages equaling 12.5 TB (2.5 TB compressed). It is hard to handle collections of the size of ClueWeb09 on a single machine, at least one needs to have a lot of external storage space. We ran our experiments on a small cluster of 15 machines; each machine costs about € 1000. The cluster runs Hadoop version 0.19.2 out of the box [14].

Time to code the experiment. After gaining some experience with Hadoop by having M.Sc. students doing practical assignments, we wrote the code for sequential search, and for anchor text extraction in less than a day. Table 1 gives some idea of the size of the source code compared to that of experimental

Table 1. Size of code base per system

Code base	#files	#lines	size (kb)
MapReduce anchors & search	2	350	13
Terrier 2.2.1	300	59,000	2,000
Lucene 2.9.2	1,370	283,000	9,800
Lemur/Indri 4.11	1,210	540,000	19,500

search systems. Note that this is by no means a fair comparison: The existing systems are general purpose information retrieval systems including a lot of functionality, whereas the linear search system only knows a single trick. The table also ignores the code needed to read the web archive "warc" format used by ClueWeb09, which consists of another 4 files, 915 lines, and 30 kb of code.[1] Still, in order to adapt the systems below, one at least has to figure out what code to adapt.

Time to run the experiment. Anchor text extraction on all English documents of ClueWeb09 takes about 11 hours on our cluster. The anchor text representation contains text for about 87 % of the documents, about 400 GB in total. A subsequent TREC run using 50 queries on the anchor text representation takes less than 30 minutes. Our linear search system implements a fairly simple language model with a length prior without stemming or stop words. It achieves expected precision at 5, 10 and 20 documents retrieved of respectively 0.42, 0.39, and 0.35 (MTC method), similar to the best runs at TREC 2009 [1].

Figure 2 shows how the system scales when processing up to 5,000 queries, using random sets of queries from the TREC 2009 Million Query track. Reported times are full Hadoop job times including job setup and job cleanup averaged over three trials. Processing time increases only slightly if more queries are processed. Whereas the average processing time per query is about 35 seconds per query for 50 queries, it goes down to only 1.6 second per query for 5,000 queries. For comparison, the graph shows the performance of "Lemur-one-node", i.e., Lemur version 4.11 running on *one fourteenth* of the anchor text representation on a single machine. A distributed version of Lemur searching the full full anchor text representation would not do faster: It would be as fast as the slowest node, it would need to send results from each node to the master, and to merge the results. Lemur-one-node takes 3.3 seconds per query on average for 50 queries, and 0.44 seconds on average for 5,000 queries. The processing times for Lemur were measured after flushing the file system cache. Although Lemur cannot process queries in parallel, the system's performance benefits from receiving many queries. Lemur's performance scales sublinearly because it caches intermediate results. Still, at 5,000 queries Lemur-one-node is only 3.6 times faster than the MapReduce system. For experiments at this scale, the benefits of the full, distributed Lemur are probably negligible.

---

[1] The warc reader was kindly provided by Mark J. Hoy, Carnegie Mellon University.

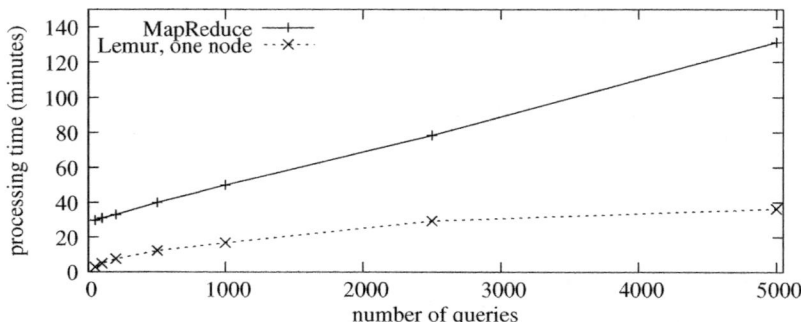

Fig. 2. Processing time for query set sizes

We also tried to index all anchor text data on a single machine and run a fraction of the queries on each of the 14 machines. Unfortunately, Lemur crashed when trying to index all data, and we did not pursue this approach any further.

On (not) computing global statistics. Our language modeling approach does not use smoothing, so it does not need global statistics to compute IDF-like (inverse document frequency) weights. Global statistics can be incorporated by doing one initial pass over the corpus to collect global statistics for all queries [2]. Post-hoc experiments using standard approaches like linear interpolation smoothing, Dirichlet prior smoothing, and Okapi's BM25 [9] show that these approaches – that *do* use global statistics – consistently perform equally or worse than our simple weighting approach. Presumably, on document collections of this scale, IDF-like weighting is unnecessary.

Related work. The idea to use sequential scanning of documents on many machines in parallel is certainly not new. Salton and Buckley [12] analyzed such a method for the Connection Machine, a multi-processor parallel computing machine. We also know of at least one researcher who used sequential scanning over ten years ago for his thesis [5]. Without high-level programming paradigms like MapReduce, however, efficiently implementing sequential scanning is not a trivial task, and without a cluster of machines the approach does not scale to large collections. A similar approach using MapReduce was taken by Lin [7], who used Hadoop MapReduce for computing pairwise document similarities. Our implementation resembles Lin's brute force algorithm that also scans document representations linearly. Our approach is simpler because our preprocessing step does not divide the collection into blocks, nor does it compute document vectors.

4 Conclusion

A faster turnaround of the experimental cycle can be achieved by making coding of experimental systems easier. Faster coding means one is able to do more experiments, and more experiments means more improvement of retrieval performance. We implemented a full experimental retrieval system with little effort using Hadoop MapReduce. Using 15 machines to search a web crawl of 0.5 billion pages, the proposed MapReduce approach is less than 10 times slower than a single node of a distributed inverted index search system on a set of 50 queries. If more queries are processed per experiment, the processing times of the two systems get even more close. The code used in our experiment is open source and available to other researchers at: `http://mirex.sourceforge.net`

Acknowledgments

Many thanks to Sietse ten Hoeve, Guido van der Zanden, and Michael Meijer for early implementations of the system. The research was partly funded by the Netherlands Organization for Scientific Research, NWO, grant 639.022.809. We are grateful to Yahoo Research, Barcelona, for sponsoring our cluster.

References

1. Clarke, C.L.A., Craswell, N., Soboroff, I.: Overview of the TREC 2009 web track. In: Proceedings of the 18th Text REtrieval Conference, TREC (2009)
2. Craswell, N., Fetterly, D., Najork, M., Robertson, S., Yilmaz, E.: Microsoft Research at TREC 2009: Web and relevance feedback tracks. In: Proceedings of the 18th Text REtrieval Conference, TREC (2009)
3. Dean, J.: Challenges in building large-scale information retrieval systems. In: Proceedings of the 2nd Conference on Web Search and Data Mining, WSDM (2009)
4. Dean, J., Ghemawat, S.: MapReduce: Simplified data processing on large clusters. In: Proceedings of the 6th Symposium on Operating System Design and Implemention, OSDI (2004)
5. Hiemstra, D.: Using Language Models for Information Retrieval. Ph.D. thesis (2001)
6. Lemur Toolkit, `http://www.lemurproject.org/`
7. Lin, J.: Brute Force and Indexed Approaches to Pairwise Document Similarity Comparisons with MapReduce. In: Proceedings of the 32nd international ACM SIGIR conference on Research and development in information retrieval (2009)
8. Lucene Search Engine, `http://lucene.apache.org`
9. Manning, C.D., Raghavan, P., Schütze, H.: Introduction to Information Retrieval. Cambridge University Press, Cambridge (2008)
10. Peters, C., Deselaers, T., Ferro, N., Gonzalo, J., Jones, G.J.F., Kurimo, M., Mandl, T., Peñas, A., Petras, V. (eds.): Evaluating Systems for Multilingual and Multimodal Information Access. LNCS, vol. 5706. Springer, Heidelberg (2009)
11. Terrier IR Platform, `http://ir.dcs.gla.ac.uk/terrier/`
12. Salton, G., Buckley, C.: Parallel text search methods. Communications of the ACM 31(2) (1988)
13. Voorhees, E.M., Harman, D.K. (eds.): TREC Experiment and Evaluation in Information Retrieval. MIT Press, Cambridge (2008)
14. White, T.: Hadoop: The Definitive Guide. O'Reilly Media, Sebastopol (2009)
15. Yu, Y., Isard, M., Fetterly, D., Budiu, M., Erlingsson, U., Kumar, P., Currey, J.: DryadLINQ: A system for general-purpose distributed data-parallel computing using a high-level language. In: Proceedings of the 8th Symposium on Operating System Design and Implementation, OSDI (2008)

Which Log for Which Information?
Gathering Multilingual Data from Different Log File Types

Maria Gäde, Vivien Petras, and Juliane Stiller

Berlin School of Library and Information Science, Humboldt-Universität zu Berlin
Dorotheenstr. 26, 10117 Berlin, Germany
{maria.gaede,vivien.petras,juliane.stiller}@ibi.hu-berlin.de
http://www.ibi.hu-berlin.de

Abstract. In this paper, a comparative analysis of different log file types and their potential for gathering information about user behavior in a multilingual information system is presented. It starts with a discussion of potential questions to be answered in order to form an appropriate view of user needs and requirements in a multilingual information environment and the possibilities of gaining this information from log files. Based on actual examples from the Europeana portal, we compare and contrast different types of log files and the information gleaned from them. We then present the Europeana Clickstream Logger, which logs and gathers extended information on user behavior, and show first examples of the data collection possibilities.

1 Introduction

In order to provide a useful and effective search and retrieval service to users, information systems need to understand who their users are, what they are looking for and how they expect to find it. Adaptive information systems change their interfaces and functionality features according to new user requirements or observed user behavior. User expectations and behavior can be gleaned from extensive questioning (interviews, surveys) or from observation of user interactions with the system (log file analysis). While qualitative methods like interviews can gather a lot of interpretive information (why do user behave that way), they can only survey a small part of the user population and can be misleading depending on the subjective notions of the questioned users. Quantitative data like log files are able to give an overview of all users active in a system. However, one might not be able to explain every observed event. Log files can be helpful in providing insights in trends and common activities as well as problems within an information system. This is valuable in exploratory analyses and can be extended and intensified with qualitative studies after a basis for questions has been established. A deep and careful evaluation of log file data can reveal more insights into user actions and interpretations of pathways through an information system than is visible at first glance, especially if the different log resources are combined with domain and system knowledge.

M. Agosti et al. (Eds.): CLEF 2010, LNCS 6360, pp. 70–81, 2010.

In this paper, we explore what kind of questions can be answered by analyzing data from log files and what data should be logged in order to understand user behavior. Especially in Europe, information systems increasingly deal with multilingual content and multilingual users alike so that questions of multilingual search capabilities and multilingual representation (starting with how many language versions of a static homepage should be created) need to be answered. We are looking at digital library applications, and in particular the search process, in order to elaborate the system components that are affected by multilingual issues and identify those aspects where data from log files can provide insight into user needs and user expectations. Our use case is Europeana[1], the European digital library, archive and museum portal with the particular charter to serve all European citizens (and their respective languages) effectively.

The remainder of the paper is organized as follows: section 2 briefly introduces Europeana, whereas section 3 reviews related work. Section 4 develops a list of questions concerning multilingual access categorized by their occurrence in the search process. Section 5 reviews common logging approaches and their outcomes for a digital library (http access logs by the web server, the search engine logs and Google Analytics log data), which are also used in Europeana, and their effectiveness in answering the questions laid out in section 4. Section 6 introduces a compound approach for logging developed for the Europeana search portal, the Europeana Clickstream Logger (CSL), which provides improved logging capabilities for answering some of the multilingual questions that could not be answered before. We conclude the paper by showing some examples of the improved logging and discussing some of the benefits of this compound approach.

2   Europeana

The Europeana portal integrates digital objects from Europe's cultural institutions such as libraries, audio-visual archives, museums and archives. The aim is to provide universal access to Europe's cultural heritage in a single access point. The material aggregated in Europeana ranges from videos and images to all kind of texts and even sounds. By the end of 2010, Europeana wants to give access to 10 million objects from all over Europe.

Europeana wants to offer access to cultural and scientific heritage material in a multilingual environment [1]. That means to overcome the language diversity in Europe and to offer access to objects in different languages regardless of the users' language skills.

Generally, multilingual access to content in digital libraries can be achieved on different levels:

1. Interface - providing parallel interfaces in different languages,
2. Browsing capabilities - providing parallel browsing structures, e.g. for a subject classification, in different languages,

---

[1] http://www.europeana.eu/

3. Search capabilities - enabling the user to find documents in a language different from the query language, and

4. Result representation - offering the possibility to translate the results in the users' preferred languages.

Europeana already displays all static interface pages in 26 European languages and provides language-sensitive search suggestions ("People are currently thinking about") from its homepage. Search results can be filtered by the language of the objects in the result list. The EuropeanaConnect project[2] is currently working on developing multilingual search (query translation) and browsing capabilities for up to 10 European languages. The development of these features is accompanied by user studies and evaluation efforts targeted towards providing further insights into user behavior (partly language-dependent) and user requirements for multilingual features within the system. Log file analyses play an important role in gathering this data.

3   Log File Studies in a Multilingual Environment

Various studies have used information from query logs to learn about the search process. Different approaches have been applied to log file analyses [2]. Previous research on query logs from search engines such as Altavista or Excite looked on general statistics, including average number of words per query, time length of query sessions, reformulation and topics [3][4]. Several studies discuss the use of clickthrough data for adequate and continuous analysis of user behavior and search engine optimizing [5]. However, very few studies seem to have targeted multilingual aspects in particular.

The Cross Language Evaluation Forum's (CLEF) iCLEF track 2008 proposed a task searching images in a naturally multilingual database (Flickr). The search logs from the Flickling search interface were analyzed with regard to user search behavior, strategies and success according to language skills [6]. In 2009, the LogCLEF track was launched with the aim to analyze and classify user queries in order to understand search behavior in multilingual contexts and to improve search systems. It consisted of two tasks: LAGI (Log Analysis and Geographic Query Identification) and LADS (Log Analysis for Digital Societies)[7]. To understand the search process and the user interaction with the system, data from The European Library (TEL) and Tumba! was evaluated. Lamm et al. [8] investigate user search performance and interaction with the TEL interface. They discovered different search behaviors of users from different countries. Bosca and Dini [9] present an approach for translation improvement by analyzing user input. The European Digital Library (TEL) and the EDLproject also conducted several log file analyses to gather user requirements. Although the full translation of documents is not required, subject keyword translation already seems to be useful for the evaluation of result relevance [10].

---

[2] http://www.europeanaconnect.eu/

The University of Padua analyzed the http traffic logs of The European Library portal finding that the majority of visitors to the portal did not perform any query. The search sessions mostly (77.44%) involved only 1 query, which are short and not reformulated, one of the main challenges regarding language detection [11]. The Max Planck Institute for Informatics in Saarbrücken analyzed server logs to research the user interaction behavior. In particular, they focused on the query and result-click history. They found that the majority of users (84%) leaves the default interface language in English [11]. The evaluation of queries showed that the most frequent searches relate to European place names or subjects.Through a study of library catalog search logs, the CACAO project found that in a library operating in a multicultural context, about 20% of the queries were written in three languages, namely Italian, German and English [12].

Previous log file studies either incorporated multilingual aspects only tangentially or in an exploratory way. This paper attempts to define a set of criteria that could be analyzed within a multilingual information environment and describes a clickstream logging approach to gather extended information on user interaction by using the example of the Europeana Clickstream Logger.

4   Logging Multilingual Information

A standard http access log file tracks every activity performed by a web server, usually page requests. From very few lines of logged data, a lot of information can be gleaned, for example, what kind of questions are asked or how long users stay on a site. Below, a set of criteria is listed, which can provide context for multilingual aspects of user behavior and system features, that could potentially be collected from observing activities logged by the system in some form.

The criteria are grouped in order of possible occurrence of the logged activity in a "classic" search process, where a user approaches a system, launches one or several queries and reviews the results.

User background information (e.g. country of access system; system language, referrer site): Looking at static system parameters like the language of the user's operating system can provide insight into the native or preferred language of the user or other routine language uses, which can be used to infer default languages for the information system's static pages and dynamic features.

Change of interface language: The change of the interface language of an information system is an active intervention of the user, possibly indicating even stronger preferences for favored languages, the incapability of understanding the default language or a stronger engagement with the system (enough interest to adapt the system to one's own uses) than the background information.

Query language: The query language indicates the language a user feels most comfortable searching in. Identifying the language of the query is crucial for the search process: language-specific query processing and indexing can lead to better search results (both in terms of ranking and in terms of serving the user with documents they can understand). If the query language is not indicated by the user, automatically detecting the query language is necessary. This can be done by analyzing the query terms, the languages of the documents found by the query or by inference from the background information like system or interface language. Sometimes, however, users change their query language in order to adapt to what they assume the language of the documents in the collection to be. Differences between the query language and the system or interface language can indicate such a switch, to which an information system can react accordingly (by suggesting different search strategies, for example). Verifying the query language through user intervention is a possible way of integrating a learning mechanism in the system although this requires careful interaction design so that the usability of the system is not hampered.

Query type (e.g. simple or default query, advanced or fielded query, related items, pre-selected categories for browsing): Different query types can indicate different user preferences or pathways through the system. The most obvious one is when a user selects a language filter for the requested results from the advanced query interface. A related item search can point towards similar documents both thematically but also in the same language. Searchers using pre-selected categories or suggested searches (like the "People are currently thinking about" feature in Europeana) might approach the system differently, not with a fixed information need in mind but serendipitous interest, to which the system can exhibit different behaviors.

Query content (e.g. named entities, dates & numbers, themes or topics): Identifying the content of the query or the object of the user request can have a profound impact on multilingual search quality. Named entities, dates and numbers, for example, should be treated differently in query translation and indexing (sometimes different language versions exist, sometimes they do not). Thematic or topical questions could be expanded or translated with the support of multilingual subject schemes like thesauri or classifications. Context- and language-sensitive recommendation features can suggest new search paths or strategies based on the initial query content.

Query translation: Query translation can be done automatically by the information system or could be offered as an additional service to the user. Logging the languages a query is translated into can provide insights into user preferences and needed feature improvements (e.g. add more languages to the translation feature). Log file data could also help in detecting different language versions of the same query or even adding unknown terms to a dictionary. If the information system offers a manual query translation feature, users can be induced to implicitly improve the translation process by suggesting more precise translations if the original and new query languages are known.

Search results: The language of the search results - if it is identified - cannot only help in query language detection but can also guide the user interaction towards restricting or expanding the result set by language or by offering value-added services like result translation if results language and other languages (interface, query) differ. Log files can give insights into the language of the particular fields within a metadata record (records themselves can contain multilingual information) that were found by a particular search or the distribution of languages over the whole results set or just the top number of search results.

Result set views: Result set views include any action the searcher performs on a result list, like opening up a particular record for a full view, paging through a result list (if it spans more than one page), saved items or even saved searches. These actions usually indicate a stronger interest of the searcher and lead to valuable information about preferred documents for a search or preferred post-search interaction scenarios.

Result translation (e.g. subjects, key metadata fields, full document): Most information systems do not offer ready-made translation capabilities for search results yet. However, user studies have shown that even translation of small but content-heavy parts of a document (e.g. the title) can support the user in deciding whether a document is relevant for their information need. Logging translation requests, the requested languages as well as views of the translations and subsequent actions can provide valuable insight into feature improvement for future multilingual development (for example, which languages are most often requested, which parts of a document should be offered for translation etc.).

User-generated content (e.g. tags): In certain information systems, users can add content to an existing metadata record; in other systems, users create a wholly new record themselves. The language of user-generated content can play an important role in observing user communities (do users describe objects in the same way) or can even be used for translation or multilingual search (especially if user-generated content in other languages than the original language of the existing metadata is added).

Query reformulation / query expansion / query refinement: After an initial search, users might replace, expand or refine the original query. If they switch languages within a session, this could indicate an attempted translation of queries to increase recall. Query reformulation can indicate language- or context-specific equivalents. Of particular interest are filtering activities, a direct intervention by the user that restricts a search by language or content provider - both indicators of language-dependent user preferences or behavior. A related item search ("more like this") is also a query reformulation since the original query is now changed based on the targeted result object. Both the path through the system and the new results can give new language-specific information.

5   Comparing Log Data

The content of a web server log depends on the server and its settings. The log entries can be presented in different forms. We briefly compare 3 types of logging data gathered at different places: the http access logs from a web server, the search engine logs and the data provided by Google Analytics. The examples are drawn from the particular case of Europeana, but could be applied to other similar digital libraries as well.

5.1   Transaction Log - Web Server

Each web server keeps log files on the transactions between server and user. Commonly, two files are written. The error log file lists the reported errors and cannot be customized; the access log can be modified to adapt to information needed. Transaction log analysis (TLA) focuses on system performance, information structure or user interaction [13].

For Europeana, Apache web server logs[3] are used. Figure 1 is an example showing a request for an image file from the image cache from a results list for the query "italy".

```
123.123.123.123 - - [11/Mar/2010:09:42:06 +0100]
"GET /cache/image/?uri=http://images.scran.ac.uk/rb/images/
thumb/0098/00980252.jpg&size=BRIEF_DOC&type=IMAGE HTTP/1.0"
200 2843 "http://www.europeana.eu/portal/brief-doc.html?
start=1&view=table&query=italy" "Mozilla/5.0 (Windows; U;
Windows NT 5.1; it; rv:1.9.2) Gecko/20100115 Firefox/3.6
(.NET CLR 3.5.30729)"
```

Fig. 1. Example line from the Europeana Apache log

The IP-address and the system language indicate the country of origin or the language (Italian) of the user. Keeping in mind that proxy use or default settings might be misleading, final conclusions about the user's language skills cannot be drawn. The IP-address, however, also implies the reach of the service and usage in different countries. The user agent requesting pages from the server is also stored in the log files. This can be used to differentiate between search engine bots crawling the site and real user sessions.

From the logs, one can also observe the interaction of the user with the system. In this example, the page requested is introduced by the "GET"-request method which is followed by the desired file, an image. Subsequent to the http status code and the number of bytes transferred is the referrer URL which shows the page visited before requesting the present page. In this example, the referrer URL shows the inital query "italy". The query can be analyzed in terms of content, language and type (advanced or simple).

Generally, it is possible to reconstruct individual user sessions, assuming the same IP address is used, and identify query reformulations and search results clicked. This requires an analysis of several log entries and the relationships between them. Multilingual information can be gleaned from the IP address, system language indicators, and the query (although only after one would go through a language detection process). However, the interface language cannot be logged this way nor the search results.

The use of Web 2.0 techniques for dynamic functionality, which is prevalent on the web today also causes some problems. These techniques make use of client-side technologies that do not necessarily communicate with the server to

---

[3] http://httpd.apache.org/

process user requests. For example, actions executed through javascript (AJAX) and Flash are often not logged in the transaction logs. For every access request, including all the style (CSS), javascript and image files are logged as well. Considering that an average html web page consist of at least 10 requests, the crucial information is easily obscured. In addition, the application state (what happens on the server between when the request arrives and the response sent back to the user) is not saved either.

5.2   Query Log - Search Engine

The query log of the search engine might only log a subset of actions the user performs on the system, namely the searches. These can also be reconstructed from the web server access logs as shown in the previous example. In the Europeana portal, the search engine used is Solr[4]. Europeana uses a stateless design, where queries are transformed into Solr syntax and then sent from the front-end servers. Because of this, session and user information is lost and cannot be reconstructed, even though the query content and query fields are contained.

The Solr search engine logs - at least for Europeana - can not add any new information that the web server log does not contain already.

5.3   Page Tagging - Google Analytics

Page tagging (adding JavaScript to a web page to gather information about access) generates detailed statistics about the use of a Web site without web server access. A prominent example, which is also used for Europeana, is Google Analytics[5]. It offers general data analysis but also provides customized reports on access statistics and strategies implemented to direct traffic streams.

Dynamic tracking allows to identify and classify the site visitors with respect to individual regions, states/provinces and cities. It is very simple to generate reports showing distributions by language (the user system language) and country or referrer sites. However, for search-based systems like Europeana, Google Analytics cannot compile information regarding interface languages (application states) or the search process (queries, results) itself.

5.4   Log File Trade-Offs

The analysis of log files comes with challenges and limitations. It is nearly impossible to identify individual users with absolute accuracy. The same user may use several IP addresses or several users can share one IP address. Furthermore, it is possible to hide the true location by using proxies. The IP address of users or any other background information can only give indications about the language preferences of users but might not be reliable in every case.

Google Analytics uses cookies in order to track visitor activity. Users who delete their cookies will still be tracked by Google Analytics, but they will be

---

[4] http://lucene.apache.org/solr/
[5] http://www.google.com/analytics/

identified as a new visitor to the site leading to incorrect session results. Using host names to group or locate users geographically can also be misleading. It is difficult to interpret user intentions from queries and search activities alone. Log entries are limited to transactions of the web server and do not reveal backgrounds or preferences [14]. They do not give any clues about the reasons behind certain actions the users executed [15]. Motives for certain actions can be very complex and dependent on the users' perception on the performance of the system.

Common log file entries are general and therefore contain limited information concerning multilingual issues. The logs cannot be used for statements about the frequency of interface language change or relation between interface language and query language. The reconstruction of individual user sessions is difficult and time-consuming. The calculation of the distribution of query or result languages cannot be easily done. Information about application states is also obscured. Information about the user and the results presented or certain client-side technology-based activities is not retrievable. For example, whether the user is logged in, how many results are returned, and statistics from which languages the results came from.

6   Europeana Clickstream Logging

Clickstream logging is a logging approach, which enables to mine complex data in order to analyze user behavior. The term "clickstream" describes the path a user takes through a website. A clickstream is a series of actions or requests on the web site accompanied by information on the activity being performed. Clickstream logging allows to track application state changes and therefore traces user behavior in a way that a traditional http transaction log is unable to. In this section, we describe a clickstream logger developed for Europeana, which pays particular attention to multilingual aspects of the portal and its users. Logging data from the Europeana Clickstream Logger (CSL) allows to reconstruct user sessions easily and provides a more complete picture on language issues.

6.1   Logging Multilingual Information in Europeana

For the Europeana clickstream logs, six different activity types or states with a particular focus on multilingual access aspects are logged. These actions indicate a stream of user activities which can be categorized as follows:

Interface language-specific actions: The interface language or the change of the interface language is logged for each transaction. This would otherwise not appear in the http access log.

Search-related actions: All search-related activities including information about query terms, result numbers and distribution of results by language and country are logged. Filtering (e.g restricting by language, provider, date) and related searches (from an initial result list) are also logged.

Browse-related actions: For the http transaction log, a browsing activity (e.g. clicking on one of the images cycling across the Europeana homepage that are suggested as search entries) is the same as a search via term entry: in both cases requests are sent to the search engine. However, from a user interaction perspective, browsing and searching might point to different user intentions. The clickstream logger logs all browsing activities and their initial starting points (e.g. did the search originate from the cycling images, the suggested searches, the time line, the saved searches links or saved user tags).

Ajax-related actions: Client-side activities (post actions) that might not appear in the transaction log, which involve interactive features of Europeana, are logged here. This includes saving or removing social tags, searches or objects.

Navigation-related actions: User paths through search results are logged here, for example, when a user moves away from Europeana by following a link from a detailed results page to the original object or when the user returns to the results list.

User management-related actions: This involves actions connected to user account creation, logging in and out and changing passwords.
Additionally, errors and requests on static pages are logged.

6.2 Europeana Clickstream Logger (CSL) Data

The comparison of a web server access log entry and the associated clickstream log shows that the clickstream approach includes much more information especially on multilingual usage. Figures 2 and 3 represent a single user action: restricting a result set by language ("fr" - french) following an initial one-word query ("treasure").
The http access log contains the query and the language facet:

```
http :// www . europeana . eu / portal / brief - doc . html ? query = treasure
& qf = LANGUAGE : fr & view = table
```

Fig. 2. Abbreviated example line from the Europeana http access log

The clickstream log shows a logged-in user (ID: 12345), who performed a search ("treasure") in the simple search interface, refined it by the language facet "fr" and now looks at the general result list. Also logged is the position in the result set (page 1), the number of returned objects (3), the result distribution by the top 5 languages and by country of the provider (top 5) and the interface language (Norwegian).

```
[action=BRIEF_RESULT, view=brief-doc-window, query=treasure,
queryType=simple, queryConstraints="LANGUAGE:"fr"", page=1,
numFound=3, langFacet=en (1,126), es (28), fr (3), mul (3),
de (2), countryFacet=france (3), userId=12345, lang=NO,
req=http://www.europeana.eu/portal/brief-doc.html?
query=treasure&qf=LANGUAGE:fr&view=table
```

Fig. 3. Abbreviated example from the Europenana CSL with action: language filter

The URL of the requested page (where the query could in this case also be reconstructed from), the referrer page, the date, time and IP address and system information for the user are noted in both log variants and are not shown in the examples.

Figure 4 shows the clickstream log entry for an interface language change - an action that cannot be logged with the http transaction log.

```
[action=LANGUAGE_CHANGE, oldLang=EN, userId=, lang=FR,
req=http://europeana.eu:80/portal/aboutus.html?lang=fr
```

Fig. 4. Abbreviated example from the Europeana CSL with action: language change

7   Conclusion

In this paper, we compare logging methods with respect to their ability to answer questions about user behavior and user intentions in a multilingual information environment. The Europeana Clickstream Logger presented above alleviates some of the information gaps that occur in traditional http access logs. The Europeana Clickstream Logger logs actions that are impossible to reconstruct from transaction logs (e.g. language changes or differences between browse and search requests), but also logs actions and their results in a way that is simpler and less error-prone to analyze. It therefore makes knowledge explicit that was previously obscured by parameters that could only be interpreted by aggregating data from different resources or that require extensive system expertise. This detailed logging enables us to gather a more complete view of the multilingual issues facing a system and to glean user intentions more explicitly. Other important multilingual aspects that cannot be studied from log files - the automatic detection of the query language, for example - still need to be tackled. However, with the clickstream logging in place, we can track how changes from the baseline (e.g. rerouting requests by language) will affect user behavior giving us an important evaluation tool for further interaction research.

Acknowledgment. Work in this paper was partially funded by the eContentplus project EuropeanaConnect (ECP-2008-DILI-528001). We would like to thank Sjoerd Siebinga from the Europeana Foundation for implementing the Europeana Clickstream Logger, his continuous support and great patience in answering our questions.

References

1. Purday, J.: Think culture: Europeana.eu from concept to construction. The Electronic Library 6, 919–937 (2009)
2. Jansen, B.J.: Search log analysis: What it is, what's been done, how to do it. Library & Information Science Research 28, 407–432 (2006)
3. Silverstein, C., Marais, H., Henzinger, M., Moricz, M.: Analysis of a very large web search engine query log. SIGIR Forum 33, 6–12 (1999)
4. Jansen, B.J., Spink, A., Bateman, J., Saracevic, T.: Real life information retrieval: a study of user queries on the web. SIGIR Forum 32, 5–17 (1998)
5. Joachims, T.: Optimizing search engines using clickthrough data. In: KDD 2002: Proceedings of the Eighth ACM SIGKDD International Conference on Knowledge Discovery and Data Mining, pp. 133–142. ACM, New York (2002)
6. Gonzalo, J., Clough, P., Karlgren, J.: Overview of iclef 2008: Search log analysis for multilingual image retrieval. In: Peters, C., Deselaers, T., Ferro, N., Gonzalo, J., Jones, G.J.F., Kurimo, M., Mandl, T., Peñas, A., Petras, V. (eds.) CLEF 2007. LNCS, vol. 5706, pp. 227–235. Springer, Heidelberg (2009)
7. Mandl, T., Agosti, M., Nunzio Di, G., Yeh, E., Mani, I., Doran, C., Schulz, J.M.: Logclef 2009: the clef 2009 multilingual logfile analysis track overview. In: Working Notes of the Cross Language Evaluation Forum, CLEF (2009)
8. Lamm, K., Mandl, T., Kölle, R.: Search path visualization and session performance evaluation with log files from the european library. In: Working Notes of the Cross Language Evaluation Forum, CLEF (2009)
9. Bosca, A., Dini, L.: Cacao project at the logclef track. In: Working Notes of the Cross Language Evaluation Forum, CLEF (2009)
10. Mane, L.: D3.2 improving full-text search in printed digital libraries' collections through semantic and multilingual functionalities - technologies assessment & user requirements. Technical report, TELplus (2009)
11. Angelaki, G.: M1.4 interim report on usability developments in the european library. Technical report, EDLproject (2007)
12. Trojahn, C., Siciliano, L.: D7.4 user requirements for advanced features. Technical report, CACAO project (2009)
13. Jansen, B.J.: The methodology of search log analysis. In: Jansen, B.J., Spink, A., Taksa, I. (eds.) Handbook of Research on Web Log Analysis. Information Science Reference, pp. 100–123 (2009)
14. Booth, D.: A review of methodologies for analyzing websites. In: Jansen, B.J., Spink, A., Taksa, I. (eds.) Handbook of Research on Web Log Analysis. Information Science Reference, pp. 143–164 (2009)
15. Hulshof, C.: Log file analysis. In: Encyclopedia of Social Measurement, vol. 2, pp. 577–583. Elsevier, Amsterdam (2004)

Examining the Robustness of Evaluation Metrics for Patent Retrieval with Incomplete Relevance Judgements

Walid Magdy and Gareth J.F. Jones

Centre for Next Generation Localization
School of Computing
Dublin City University, Dublin 9, Ireland
{wmagdy,gjones}@computing.dcu.ie

Abstract. Recent years have seen a growing interest in research into patent retrieval. One of the key issues in conducting information retrieval (IR) research
is meaningful evaluation of the effectiveness of the retrieval techniques applied
to task under investigation. Unlike many existing well explored IR tasks where
the focus is on achieving high retrieval precision, patent retrieval is to a significant degree a recall focused task. The standard evaluation metric used for patent
retrieval evaluation tasks is currently mean average precision (MAP). However
this does not reflect system recall well. Meanwhile, the alternative of using the
standard recall measure does not reflect user search effort, which is a significant
factor in practical patent search environments. In recent work we introduce a
novel evaluation metric for patent retrieval evaluation (PRES) [13]. This is designed to reflect both system recall and user effort. Analysis of PRES demonstrated its greater effectiveness in evaluating recall-oriented applications than
standard MAP and Recall. One dimension of the evaluation of patent retrieval
which has not previously been studied is the effect on reliability of the evaluation metrics when relevance judgements are incomplete. We provide a study
comparing the behaviour of PRES against the standard MAP and Recall metrics
for varying incomplete judgements in patent retrieval. Experiments carried out
using runs from the CLEF-IP 2009 datasets show that PRES and Recall are
more robust than MAP for incomplete relevance sets for this task with a small
preference to PRES as the most robust evaluation metric for patent retrieval
with respect to the completeness of the relevance set.

1 Introduction

Interest in patent retrieval research has shown considerable growth in recent years.
Reflecting this patent retrieval has been introduced as a task at two of the major information retrieval (IR) evaluation campaigns NTCIR and CLEF in 2003 and 2009
respectively. The aim of these tasks at these workshops is to encourage researchers to
explore patent retrieval on common tasks in order understand the issues in providing
effective patent retrieval and to establish the best IR methods for doing this. Due to
the important of finding all items relevant to the query, patent retrieval is generally
identified as a recall-oriented retrieval task [7]. While in practice achieving 100%

M. Agosti et al. (Eds.): CLEF 2010, LNCS 6360, pp. 82–93, 2010.
© Springer-Verlag Berlin Heidelberg 2010

recall may not be achievable, the operational objective of patent retrieval is to maximise search recall for the minimum effort by the user, typically a professional patent examiner. In contrast to this, the majority of IR evaluation tasks are precision focused IR tasks, where one or two of the relevant documents are often sufficient to achieve user satisfaction by addressing their information need. In this latter case the objective is to find these relevant documents as quickly as possible at high rank in a retrieved document list to minimise user effort. While considerable effort has been devoted to the study of evaluation metrics for precision focused IR tasks, the evaluation of patent retrieval is still an open area of research due to the special objective of the search task. The standard evaluation metric used for most IR tasks remains mean average precision (MAP). While patent retrieval is recognised to be a recall-oriented IR task, MAP is still the most widely used score for patent retrieval evaluation. In previous work we conducted a careful examination of the suitability of MAP for evaluating patent retrieval [13]. As a result of this investigation we introduced a new patent retrieval evaluation score (PRES) as a new evaluation metric specifically designed for recall-oriented IR tasks.

Laboratory evaluation for IR tasks such as those examined at TREC, CLEF and NTCIR relies on a standard model of a target document collection, representative search topics and a relevance set indicating which documents from the document set are relevant to each topic. An important practical issue is that the size of realistic document collections means not all documents can be assessed for relevance to each topic. An approximation to the true complete relevance is made based on some reasonable method for example using a pooling procedure involving manual relevance assessment of a selected subset of the document set or by assuming a relevance relationship between the topic and a subset of the documents [6]. In consequence the set of known relevant items for each topic is almost certainly incomplete. An important question which has been examined in a number of studies is the extent to which experimental results for IR tasks are robust to the design of the test collections [3, 4, 17]. One aspect of this research has focused on the number of topics to be used for an IR experiment to achieve reliable and robust results [4, 17], while others have focused on how evaluation metrics are stable with different values for cut-off levels of the retrieved results list [4]. Other work has studied the robustness of evaluation scores relating to the incompleteness of relevance judgements [2]. This latter study examined the stability of ranking system effectiveness when using different portions of the relevance judgements. Some score metrics have been developed to overcome problems relating to incomplete relevance assessment, such as inferred average precision (infAP) [1] and Bpref [5]. Although these metrics have proven some robustness to incomplete judgments, they are focused on the precision of retrieval results, which is not the objective of the recall-oriented patent retrieval task. An important and previously unexplored issue is the examination of the behaviour of evaluation metrics with incomplete judgements studies when evaluating a recall-oriented IR application. Similarly to standard precision-orientated tasks, such as ad hoc search, relevance sets in patent retrieval will generally be incomplete. Thus an important question is the stability of evaluation metrics across different IR systems used for patent search variations in the incomplete relevance set. In the case of patent retrieval, the practical significance of this is to assess the extent to which the value of evaluation metric measured using an incomplete relevance set is a prediction of its ability to help the patent examiner to find further relevant documents if they continued to search in the collection.

This paper provides what we believe to be the first study of the robustness of evaluation metrics for patent retrieval with incomplete relevance judgements. We compare the behaviour of standard MAP and Recall and with our PRES metric designed for the patent retrieval task. We compare behaviour of these three metrics since the first two are the ones most commonly used to evaluate effectiveness of a patent retrieval system, and the third is specifically designed for evaluation of patent retrieval and has shown desirable behaviour in terms of evaluating submissions to the CLEF-IP task [13]. Our investigation uses similar techniques to those described in [3] to test the robustness of these three metrics, although some modification is required due to the small number of relevant documents per topic. Experiments were performed on 48 runs submitted by participants in the CLEF-IP 2009 task [16]. Results show strong robustness for PRES and Recall for this type of task, while MAP is shown to have much weaker robustness for MAP to variations in the known relevance set. Our conclusion is that these results indicate that MAP is not a suitable evaluation metric for recall-oriented IR tasks in general and patent retrieval in particular, especially in the absence of a guarantee of the completeness of the relevance judgements.

The remainder of this paper is organized as follows: Section 2 provides background on patent retrieval and the evaluation metrics used for this task, Section 3 describes our experimental setup and provides relevant details of the CLEF-IP 2009 task, Section 4 describes the results of our investigation of the robustness of the evaluation scores, and finally Section 5 concludes and provides possible direction for future research.

2 Background

This section provides an introduction to the task of patent retrieval and reviews the history of its introduction in the NTCIR and CLEF IR evaluation campaigns. In addition, the evaluation metrics commonly used to evaluate patent retrieval are described in summary.

2.1 Patent Retrieval

Evaluation of patent retrieval was proposed in NTCIR-2 in 2001 [12]. Since then patent retrieval has featured as a track in all NTCIR[1] campaigns. Patent retrieval was introduced much more recently at CLEF[2] in 2009, as the CLEF-IP (CLEF Intellectual Property) track [16]. Patent retrieval is of interest in IR research since it is of commercial interest and is a challenging IR task with different characteristics to popular IR tasks such as precision-oriented ad hoc search on news archives or web document collections [12, 16]. Various tasks have been created around patents; some are IR activities while others focus on tasks such as data mining from patents and classification of patents.

---

[1] http://www.nii.ac.jp/
[2] http://www.clef-campaign.org/

The IR tasks at NTCIR and CLEF related to patent retrieval are as follows:

Ad-hoc search. A number of topics are used to search a patent collection with the objective of retrieving a ranked list of patents that are relevant to this topic [10].

Invalidity search. The claims of a patent are considered as the topics, and the objective is to search for all relevant documents (patents and others) to find whether the claim is novel or not [7]. All relevant documents are needed, since missing only one document can lead to later invalidation of the claim or the patent itself.

Passage search. The same as invalidity search, but because patents are usually long, the task focuses on indicating the important fragments in the relevant documents [8].

Prior-art search. In this task, the full patent is considered as the topic and the objective is to find all relevant patents that can invalidate the novelty of the current patent, or at least patents that have parts in common with the current patent [16]. In this type of task, patent citations are considered as the relevant documents, and the objective is to automatically find these patent citations. These citations are usually identified by a patent office and take considerable periods of time to search for them manually [9, 16].

In this paper, experiments examine prior-art search task for patent retrieval. This kind of task is characterized by the small number of relevant documents per topic. This small number of relevant items is to be expected since any filed patent should contain novel ideas that typically should not be contained in prior patents. Although the fact that these citations take a huge amount of effort and time to identify, there is no guarantee that the recall of finding all the relevant documents to be 100%[3]. Hence, the focus of the study in this paper is to examine the impact on evaluation metrics of missing any of these relevant documents.

2.2 Evaluation Metrics for Patent Retrieval

Mean Average Precision (MAP). While many evaluation metrics have been proposed for ad hoc type IR tasks, by far the most popular in general use is MAP [1]. The standard scenario for use of MAP in IR evaluation is to assume the presence of a collection of documents representative of a search task and a set of test search topics (user queries) for the task along with associated manual relevance data for each topic. For practical reasons the relevance data is necessarily not exhaustive, but the relevance data for each topic is assumed to be a sufficient proportion of the relevant documents from the document collection for this topic. "Sufficient" relates to the reliability of the relative ranking of the IR systems under examination. Several techniques are available for determining sufficient relevant documents for each topic [6]. As its name implies, MAP is a precision metric, which emphasizes returning relevant documents earlier in a ranked list (Equation 1). The impact on MAP of locating relevant documents later in the ranked list is very weak, even if very many such documents have been retrieved. Thus while MAP gives a good and intuitive means of comparing systems for IR tasks emphasising precision, it will often not give a meaningful interpretation for recall-focused tasks. Despite this observation, MAP remains the standard evaluation metric used for patent retrieval tasks.

---

[3] Based on discussions with staff from the European Patent Office.

$$Avg.\ precision = \frac{\sum_{r=1}^{N}(P(r) \times rel(r))}{n} \tag{1}$$

where r is the rank, N the number of documents retrieved, $rel(r)$ a binary function of the document relevance at a given rank, $P(r)$ is precision at a given cut-off rank r, and n is the total number of relevant documents ($|\{relevant\ documents\}|$).

Recall. Recall is a standard score metric measuring the ability of an IR system to retrieve relevant documents from within a test collection. It gives no indication of how retrieved relevant documents are ranked within the retrieved ranked list. For this reason recall is generally used in conjunction other evaluation metrics, typically precision, to give a broader perspective on the behaviour of an IR system. Although recall is the main objective in patent retrieval, it has not been used as a performance score to rank different systems due to its failure to reflect the quality of ranking of the retrieved results. Nevertheless, recall remains a very important metric to show this aspect of a patent retrieval system's behaviour which is not reflected by precision measures such as MAP.

Patent Retrieval Evaluation Score (PRES). Based on a review of the objectives of patent retrieval and the deficiencies existing metrics, in earlier work we introduced PRES as a novel metric for evaluating recall-oriented IR applications [13]. PRES is derived from the normalized recall measure (R_{norm}) [15]. It measures the ability of a system to retrieve all known relevant documents earlier in the ranked list. Unlike MAP and Recall, PRES is dependent on the relative effort exerted by users to find relevant documents. This is mapped by N_{max} (Equation 2), which is an adjustable parameter that can be set by users and indicates the maximum number of documents they are willing to check in the ranked list. PRES measures the effectiveness of ranking documents relative to the best and worst ranking cases, where the best ranking case is retrieving all relevant documents at the top of the list, and the worst is to retrieve all the relevant documents just after the maximum number of documents to be checked by the user (N_{max}). The idea behind this assumption is that getting any relevant document after N_{max} leads to it being missed by the user, and getting all relevant documents after N_{max} leads to zero Recall, which is the theoretical worst case scenario. Figure 1 shows an illustrative graph of how to calculate PRES, where PRES is the area between the actual and worst cases (A_2) divided by the area between the best and worst cases (A_1+A_2).

N_{max} introduces a new definition to the quality of ranking of relevant results, as the ranks of results are relative to the value of N_{max}. For example, getting a relevant document at rank 10 will be very good when $N_{max}=1000$, good when $N_{max}=100$, but bad when $N_{max} = 15$, and very bad when $N_{max}=10$. Systems with higher Recall can achieve a lower PRES value when compared to systems with lower Recall but better average ranking. The PRES value varies from R to nR^2/N_{max}, where R is the Recall, according to the average quality of ranking of relevant documents.

$$PRES = \frac{A_2}{A_1 + A_2} = 1 - \frac{\dfrac{\sum r_i}{n} - \dfrac{n+1}{2}}{N_{max}} \tag{2}$$

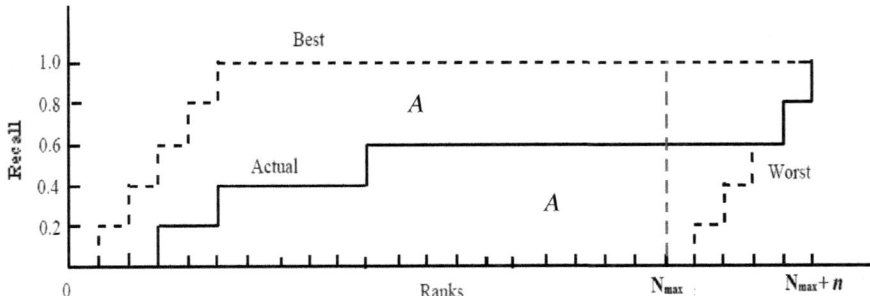

Fig. 1. PRES curve is bounded between the best case and the new defined worst case

where r_i is the rank at which the ith relevant document is retrieved, N_{max} is the maximum number of retrieved documents to be checked by the user, i.e. the cut-off number of retrieved documents, and n is the total number of relevant documents.

In [13] we demonstrate that PRES is a more suitable score metric for ranked recall-oriented IR applications than existing evaluation metrics. This was illustrated using participants' result runs submitted to the CLEF-IP 2009 which have been released for research purposes. This earlier study did not explore the robustness of the compared evaluation measures with respect to the completeness of the relevance set. The study in this paper extends our existing work to this important previously overlooked dimension of the behaviour of the metrics.

In the remainder of this paper describes our investigation is into the stability of MAP, Recall, and PRES for the same experimental data in [13] using a similar experimental approach to that used in [3].

3   Experimental Setup

The study in this paper is performed on the CLEF-IP 2009 patent retrieval task [16]. The submitted runs for the main task in CLEF-IP 2009 are used to compare the robustness of evaluation metrics when relevance judgements are incomplete.

3.1   CLEF-IP 2009 Track

The aim of the CLEF-IP track is to automatically find prior art citations for patents. The topics for this task are patents filed in the period after 2000. The collection to be searched contains about one million patents filed in the period from 1985 to 2000 [16]. The main task in the track was prior-art search (section 2.1); where the objective is to automatically retrieve all cited patents found in the collection. These citations are originally identified by the patent applicant or the patent office.

Forty-eight runs were submitted by the track participants. The track organizers have been kind enough to release these runs in order to encourage investigation of new evaluation methodologies for patent retrieval. Each run consists of up to the top 1000 ranked results for each topic. The topic set consists of 500 topics, which is the smallest topic set provided by the track. Different topic sets of sizes up to 10,000 topics were available. However the number of runs submitted for these topic sets was much less than 48. In our previous research developing PRES and comparing it to MAP and Recall with these 48 runs, only 400 topics were used in order to anonomize IDs of the runs of the individual participants [13]. The same set of runs using the 400 topic subset of each run is used in this investigation. The average number of relevant documents per topic is 6 [16] with a minimum number of 3. This is considerably less than is typically found for existing standard ad hoc IR evaluation tasks. Figure 2 shows the number of relevant documents per each topic for the 400 topic set used in the experiments. Figure 2 shows that more than 25% of the topics have only 3 relevant documents. This small number emphasizes the importance of the robustness of the evaluation metric used since missing even one relevant document can have a huge impact on the relationship between the topic and the collection.

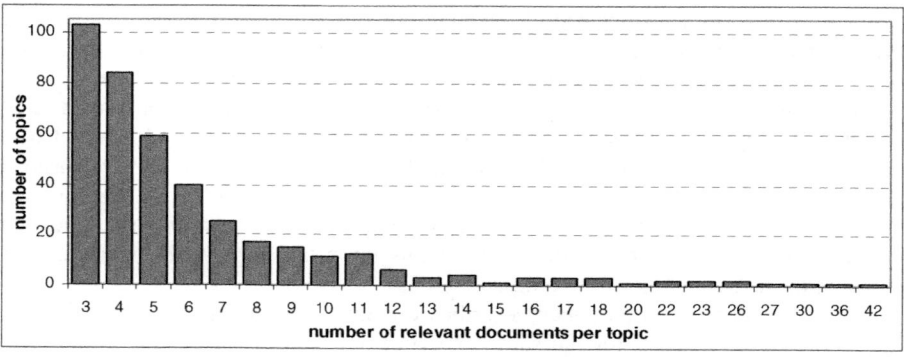

Fig. 2. Distribution of the number of relevant documents per topic in the test collection (average number of relevant documents = 6)

3.2 Experimental Setup

In order to check the robustness of the evaluation metrics, several versions of the relevance judgements *qrels* were created by generating different fractions of the judgements *f-qrels*. The original *qrels* provided by the track are assumed to be the full 100% *qrels*[4]; other versions representing fractions of the *qrels* were generated by selecting a certain fraction value (*f*) of the relevant documents for each topic in the topic set. In [3], a similar setup was used to the study in this paper, but with two main differences. The first is that this study is focusing on a recall-oriented patent retrieval task, where missing any relevant document in the assessments should be considered very harmful for a fair evaluation of systems. The second difference lies in the nature of the data collections used in the studies. In [3] TREC collections characterized by a high average number of relevant documents per topic were used. This allowed the study to test many *f-qrels*, where *f* ranged from 0.01 to 0.9. In the current experiments, patent collection is used, which is characterized by the relatively low average

---

[4] Although of course this is actually known not be the case since exhaustive manual analysis of the collection has not been carried out.

number of relevant documents per topic. This low number does not allow such large variation in the values of f.

To conduct the robustness experiments, four fraction values of the *qrels* were used (f = 0.2, 0.4, 0.6, and 0.8). For each f value, three *f-qrels* are generated, where the selection of the fraction of relevant documents was randomized, hence the three versions are always different. This produced a set of 12 *f-qrels*. The objective is to compare the ranking of the runs according to each score using these *f-qrels* to the ranking when using the full *qrels*. Kendall's tau correlation [11] was used to measure the change in the ranking. The higher the correlation for smaller values of f, the more robust the metric is to the incompleteness of relevance judgements.

Two values of cut-offs were used, the first is the one reported in the CLEF-IP track itself which is 1000 results for each topic. The second cut-off value is 100, which is more realistic for a patent retrieval task since this is the order of the number of documents typically checked for relevance by a patent examiner for each topic.

4 Results

Table 1 shows the Kendall tau correlation values for the three scores at different cut-offs and for the different samples for each value of f (% *qrels*). Figures 3 and 4 plot the worst-case values of the correlation for the three scores for cut-offs values of 100 and 1000 respectively. Table 1 and Figures 3 and 4 demonstrate several points:

- MAP has a much lower Kendall tau correlation when compared to the Recall and PRES, especially for lower values of f. This result surprisingly shows that the most commonly used metric for patent retrieval evaluation is the least reliable one when there is no guarantee of the completeness of the relevance judgements.
- Recall and PRES have nearly symmetric performance with incomplete judgements with slightly better performance to PRES for lower values of cut-off.
- Following the study of Voorhees [18] which determines rankings to be nearly equivalent if they have a Kendall tau correlation value of 0.9 or more, and to have a noticeable difference for Kendall tau correlation less than 0.8. According to the results found in our investigation, Recall and PRES will have an equivalent ranking for systems even with only 20% of the relevance judgements. However, MAP may have a noticeable change of system ranking even if only 20% of the judgements are missing.
- A drop in the curve of correlation of MAP for cut-off of 100 can be seen from Figure 3 when % *qrels* = 60%. One explanation for this can be the randomness used in selecting the fraction of relevant document from the *qrels*.

Although PRES and Recall have similar performance with the incomplete judgements, both metrics cannot be claimed to be the same. The experiments in this paper only test one aspect of the evaluation metrics. However, additional factors should be taken into consideration when considering the suitability of an evaluation metric, in this case the metric's ability to distinguish between the performances of different systems in a fair way. Bearing all factors in mind, PRES can be considered as the more

Table 1. Correlation between the ranking of 400 topics from 48 runs with different percentages of incomplete judgements and different cut-offs for MAP, Recall, and PRES

% qrels	Sample	Cut-off = 100			Cut-off = 1000		
		MAP	Recall	PRES	MAP	Recall	PRES
20%	1	0.50	0.94	0.94	0.58	0.93	0.93
	2	0.71	0.88	0.92	0.70	0.92	0.89
	3	0.59	0.90	0.90	0.66	0.90	0.90
	avg.	0.60	0.90	0.92	0.65	0.92	0.91
40%	1	0.93	0.92	0.93	0.76	0.96	0.96
	2	0.71	0.93	0.93	0.87	0.95	0.93
	3	0.75	0.91	0.92	0.84	0.94	0.92
	avg.	0.79	0.92	0.93	0.82	0.95	0.94
60%	1	0.66	0.95	0.96	0.82	0.98	0.98
	2	0.66	0.95	0.97	0.82	0.98	0.97
	3	0.65	0.96	0.96	0.90	0.97	0.98
	avg.	0.66	0.95	0.97	0.85	0.98	0.97
80%	1	0.79	0.96	0.97	0.96	0.98	0.98
	2	0.94	0.97	0.99	0.94	0.97	0.98
	3	0.75	0.98	0.98	0.84	0.98	0.99
	avg.	0.83	0.97	0.98	0.91	0.98	0.98
100%	NA	1	1	1	1	1	1

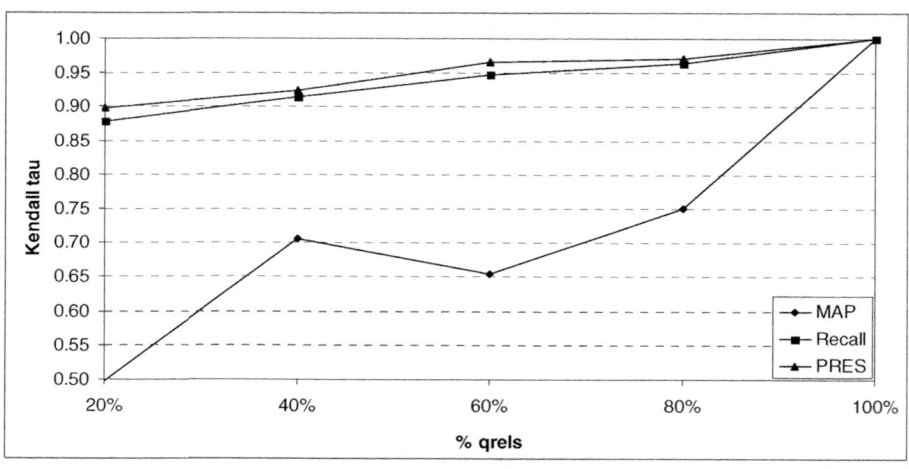

Fig. 3. Lowest Kendall tau correlation values for MAP/Recall/PRES for cut-off = 100

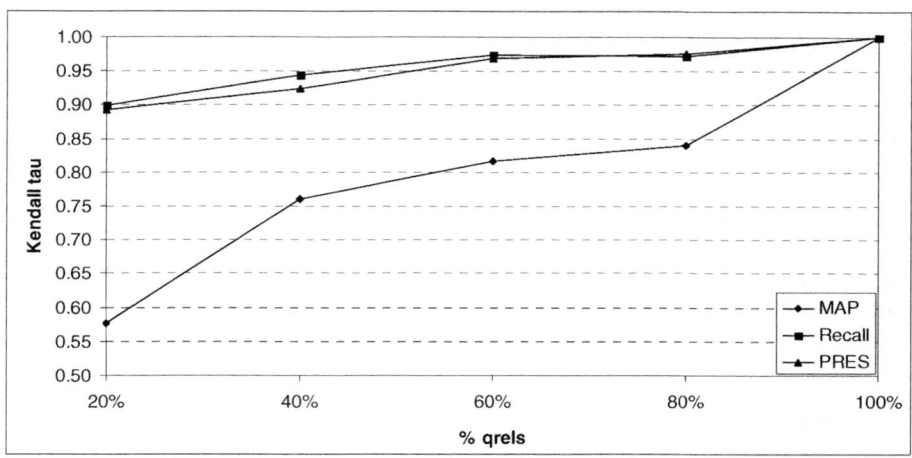

Fig. 4. Lowest Kendall tau correlation values for MAP/Recall/PRES for cut-off = 1000

suitable evaluation metric for patent retrieval since it has been shown to have a greater ability to rank systems in a recall-oriented IR environment [13]. Both this and the findings in this paper with regard to the consistent performance of PRES across different fractions of the relevance judgements recommend the use of PRES for this type of IR application.

5   Conclusion and Future Work

In this paper, a study for the robustness of the evaluation metrics used for the recall-oriented patent retrieval task has been presented. The aim of the study was to test the consistency of the performance of three evaluation metrics which are currently used for patent retrieval evaluation (MAP, Recall, and PRES) when the relevance judgement set is incomplete. Different fractional values of the *qrels* with different samples were used to conduct the experiments. Kendall tau correlation was used as the measure of the consistency of the ranking of systems. Results show that the most commonly used score for evaluating patent retrieval, MAP, is the least reliable evaluation metric to be used in this kind of IR application, since it shows the least consistency in ranking different runs when the relevance judgements are incomplete. PRES and Recall both have very robust performance even when only small portions of the relevant judgements are available. Considering the strong performance of PRES for evaluating recall-oriented IR applications, in addition to the results of this paper; PRES can be recommended as a standard score metric for evaluating recall-oriented IR application, especially patent retrieval.

As future work, this study can be extended to include more runs submitted to the upcoming CLEF-IP tracks. PRES is due to be considered as one of the metrics used for evaluating this task in CLEF-IP in 2010. It will be interesting to see whether these new runs provide further evidence for the robustness of these metrics. Furthermore, the study could be expanded to capture other types of evaluation metrics, such as geometrical mean average precision (GMAP), precision at different cut-off values, and normalized discount cumulative gain (NDCG). Although this kind of study can be interesting from the robustness point of view, there is always another dimension which needs to be considered when selecting an evaluation metric namely, what feature of a system's behaviour is the metric evaluating.

Acknowledgment

This research is supported by the Science Foundation Ireland (Grant 07/CE/I1142) as part of the Centre for Next Generation Localisation (CNGL) project at Dublin City University.

References

1. Aslam, J.A., Yilmaz, E.: Estimating average precision with incomplete and imperfect judgments. In: Proceedings of the 15th ACM international conference on Information and knowledge management CIKM, Arlington, Virginia, USA, pp. 102–111 (2006)
2. Baeza-Yates, J., Ribeiro-Neto, B.: Modern Information Retrieval. Addison-Wesley, Reading (1999)
3. Bompad, T., Chang, C.-C., Chen, J., Kumar, R., Shenoy, R.: On the robustness of relevance measures with incomplete judgements. In: Proceedings of the 30th Annual International ACM SIGIR Conference on Research and Development in Information Retrieval, Amsterdam, The Netherlands. ACM, New York (2007)
4. Buckley, C., Voorhees, E.: Evaluating evaluation measure stability. In: Proceedings of the 23rd Annual International ACM SIGIR Conference on Research and Development in Information Retrieval, pp. 33–40. ACM, New York (2000)
5. Buckley, C., Voorhees, E.M.: Retrieval evaluation with incomplete information. In: Proceedings of the 23rd Annual International ACM SIGIR Conference on Research and Development in Information Retrieval, Sheffield, South Yorkshire, UK, pp. 25–32 (2004)
6. Buckley, C., Dimmick, D., Soboroff, I., Voorhees, E.: Bias and the limits of pooling. In: Proceedings of the 29th Annual International ACM SIGIR Conference on Research and Development in Information Retrieval, Seattle, WA, USA, pp. 619–620. ACM, New York (2006)
7. Fujii, A., Iwayama, M., Kando, N.: Overview of patent retrieval task at NTCIR-4. In: Proceedings of the Fourth NTCIR Workshop on Evaluation of Information Retrieval, Automatic Text Summarization and Question Answering, Tokyo, Japan (2004)
8. Fujii, A., Iwayama, M., Kando, N.: Overview of the patent retrieval task at the NTCIR-6 workshop. In: Proceedings of the 6th NTCIR Workshop Meeting on Evaluation of Information Access Technologies: Information Retrieval, Question Answering and Cross-lingual Information Access, Tokyo, Japan, pp. 359–365 (2007)
9. Graf, E., Azzopardi, L.: A methodology for building a patent test collection for prior art search. In: Proceedings of The Second International Workshop on Evaluating Information Access (EVIA 2008), Tokyo, Japan (2008)
10. Iwayama, M., Fujii, A., Kando, N., Takano, A.: Overview of patent retrieval task at NTCIR-3. In: Proceedings of the 3rd NTCIR Workshop on Evaluation of Information Retrieval, Automatic Text Summarization and Question Answering, Tokyo, Japan (2003)
11. Kendall, M.: A new measure of rank correlation. Biometrika 30(1/2), 81–93 (1938)

12. Leong, M.K.: Patent Data for IR Research and Evaluation. In: Proceedings of the 2nd NTCIR Workshop Meeting on Evaluation of Information Access Technologies: Information Retrieval, Question Answering and Cross-lingual Information Access, Tokyo, Japan, pp. 359–365 (2001)
13. Magdy, W., Jones, G.J.F.: PRES: a score metric for evaluating recall-oriented information retrieval applications. In: Proceedings of the 33rd Annual International ACM SIGIR Conference on Research and Development in Information Retrieval, Geneva, Switzerland. ACM, New York (2010)
14. Van Rijsbergen, C.J.: Information Retrieval, 2nd edn. Butterworths (1979)
15. Robertson, S.E.: The parametric description of the retrieval tests. Part 2: Overall measures. Journal of Documentation 25(2), 93–107 (1969)
16. Roda, G., Tait, J., Piroi, F., Zenz, V.: CLEF-IP 2009: Retrieval experiments in the Intellectual Property domain. In: CLEF 2009 Working Notes, Corfu, Greece (2009)
17. Voorhees, E.M.: Special Issue: The Sixth Text REtrieval Conference (TREC-6). Information Processing and Management, 36(1) (2000)
18. Voorhees, E.M.: Evaluation by highly relevant documents. In: Proceedings of the 24th ACM SIGIR Conference on Research and Development in Information Retrieval, New Orleans, U.S.A., pp. 74–82 (2001)

On the Evaluation of Entity Profiles

Maarten de Rijke[1], Krisztian Balog[1], Toine Bogers[2], and Antal van den Bosch[3]

[1] ISLA, University of Amsterdam, The Netherlands
{derijke,k.balog}@uva.nl
[2] IIIA, Royal School of Library & Information Science, Denmark
tb@db.dk
[3] ILK/Tilburg Centre for Creative Computing, Tilburg University, The Netherlands
antal.vdnbosch@uvt.nl

Abstract. Entity profiling is the task of identifying and ranking descriptions of a given entity. The task may be viewed as one where the descriptions being sought are terms that need to be selected from a knowledge source (such as an ontology or thesaurus). In this case, entity profiling systems can be assessed by means of precision and recall values of the descriptive terms produced. However, recent evidence suggests that more sophisticated metrics are needed that go beyond mere lexical matching of system-produced descriptors against a ground truth, allowing for graded relevance and rewarding diversity in the list of descriptors returned. In this note, we motivate and propose such a metric.

1 Introduction

Entity retrieval is concerned with the identification of information relevant to information needs that concern entities (people, organizations, locations, products, ...) [4]. Entity finding systems return ranked lists of entities in response to a keyword query. Entity profiling systems return a ranked list of descriptions that (together) describe an entity. The profiling task can be viewed as a summarization or question answering type task for which a set of "information nuggets" needs to be extracted from a collection of documents [15]. Appropriate evaluation methodology has been defined, and later refined, by a number of authors; see, e.g., [13]. Alternatively, entity profiling systems can be viewed as systems that need to select a set of descriptors (from a knowledge source) that accurately describe a given input entity. E.g., when the type of entity of interest is people, the descriptors can be taken from an ontology describing a scientific discipline and the profiling system's task could be interpreted as expert profiling: for every individual, to identify the areas in which he or she is an expert [2]. This second, descriptor-based reading of entity profiling is the one on which we focus.

Evaluation of descriptor-based entity profiling is usually done in terms of precision and recall of the lists of descriptors produced by a system. This has several shortcomings. Returning a ranked list of descriptors for an entity is challenging. When descriptors are to be taken from a large knowledge source, near misses are likely. But not all mistakes are equally important, depending, in part, on the envisaged users. For users that are relatively new to the area described by the knowledge source, near misses that are too specific may be more problematic than ones that are too general; for expert

M. Agosti et al. (Eds.): CLEF 2010, LNCS 6360, pp. 94–99, 2010.

users, this may be the other way around. Also, some descriptors may be more fitting than others, thus naturally leading to graded relevance values. Finally, the presence of closely related descriptors in a result set at the cost of omitting descriptors that highlight different aspects of an entity would certainly be viewed unfavorably by all users.

Building on work on novelty and diversity in information retrieval evaluation [1, 8, 11], we develop a scoring method for entity profiles that addresses many of the shortcomings of today's prevalent evaluation method. Our method allows for weighted non-exact matches between system-produced descriptors and ground truth descriptors; it systematically rewards more highly relevant descriptors and more diverse lists.

2   Motivation

To motivate the need for more sophisticated evaluation methods than straightforward precision/recall of descriptors, we build on a recent evaluation of an expert profiling system. We focus on the topical expert profiling task and use the UvT Expert collection [6] as our experimental platform; it is based on the Webwijs ("Web wise") system developed at Tilburg University (UvT) in the Netherlands. Webwijs is a database of UvT employees who are involved in research or teaching; each expert can self-assess his/her skills by selecting expertise areas from a hierarchy of descriptors.

Recently, a university-wide experiment was conducted at UvT in which expertise profiles were automatically generated and, subsequently, each employee was given the opportunity to assess the profile that was generated for him or her. Specifically, participants were given a list of descriptors proposed by the profiling system. For each descriptor, participants were asked to indicate whether it correctly describes one of his or her areas of expertise. Optionally, for a given descriptor participants could indicate their level of expertise on scale of 1 (lowest) to 5 (highest). Finally, they could leave behind any comments they wished to share. A total of 246 people self-assessed their (automatically generated) profiles. Of these, 226 indicated their levels of expertise on a scale of 1 to 5. Also, 89 participants supplied comments on the proposed profiles. In a separate study [5], we transcribe and analyze these comments through content analysis. Rather than reproducing the outcomes here, we share a selection of assessors' comments that support our proposed evaluation framework.

First, the feedback we received from our participants does signal a need for more than simply performing lexical matching. Users higher up in the organizational hierarchy, such as full professors, tend to prefer more specific expertise descriptors. One commonly mentioned reason for this is that narrower expertise descriptors tend to streamline communication and enable interested parties to directly contact the relevant expert. A specific example of this is a professor in psychopharmacology at UvT, who did not want to select the expertise keyword 'Drugs' as an expertise area, because it would result in a whole range of questions that are not part of his expertise. In contrast, teachers and research assistants at UvT tend to prefer broader terms to describe their expertise.

In the feedback we received, only one participant doubted the usefulness of rating one's expertise; 226 (out of 246) used multiple values on the rating scale. This lends credit to the idea of using graded relevance values for describing someone's expertise.

As to the importance of diversity of recommendations, several participants signaled a need for minimizing overlap in the recommended expertise descriptors. E.g., one person complained about being recommended both 'international public law,' 'international law,' and 'international private law,' which are all near-synonyms. A profiling system that focuses more on diversity could help avoid such problems.

3   Scoring Profiles

We present our evaluation framework in five steps, using the following notation:

- d: a descriptor (i.e., label, thesaurus term, ...) that may or may not be relevant to an information need e; the output of an entity profiling system is a ranked list of descriptors $d_1, \ldots, d_k$;
- $\Delta = \{d_1, \ldots, d_m\}$: the set of all possible descriptors;
- e: an entity for which a profile is being sought (the user's information need); we model e as a set of descriptors $e \subseteq \Delta$;
- R: a binary random variable representing relevance.

We follow Clarke et al. [8] in using the probability ranking principle as the starting point for the definition of a scoring method to be used for assessing the output of an entity profiling system. Our aim, then, is to estimate $P(R = 1|e, d)$, the probability of relevance given information need e and descriptor d.

3.1   Baseline Approach

The standard way of assessing a ranked list of descriptors output by an entity profiling system is in terms of precision and recall of the descriptors retrieved [2]:

$$P(R = 1|e, d) = P(d \in e). \tag{1}$$

Traditionally, the probabilities are estimated to be 0 or 1 for particular choices of information need e and descriptor d; $P(d \in e)$ indicates that d is known to be a valid descriptor for e and $P(d \in e) = 0$ indicates that d is known not to be a valid descriptor. This traditional model only allows exact lexical matches with ground-truth descriptors.

3.2   Beyond Lexical Matching

We generalize (1) to allow for a more relaxed matching between a system-produced descriptor d and descriptors d_j contained in the ground truth: rather than requiring that $d = d_j$ (for some j), we ask that some d_j that is known to be relevant "provides support" for d. More precisely, we assume independence of $d_j \in e$ and $d_k \in e$ (for $j \neq k$) and reward absence of non-relevance:

$$P(R = 1|e, d) = 1 - \left(\prod_{j=1}^{m} (1 - P(d_j \in e) \cdot P(d|d_j)) \right). \tag{2}$$

Here, $P(d_j \in e)$ denotes the probability that e is correctly described by d_j and $P(d|d_j)$ denotes the probability that d_j supports d. We turn to $P(d_j \in e)$ in Section 3.3; for $P(d|d_j)$ there are several natural estimations. E.g., it could be corpus-based or a probabilistic semantic measure derived from the structure of Δ, the space of all descriptors, based on conceptual relationships.

3.3   Assessments

How should we estimate $P(d_j \in e)$? We adopt a model inspired by the way in which (topical) profiles are often determined for humans [2, 3]. We assume that a human assessor presented with information about entity e reaches a graded decision regarding each descriptor $d_j \in \Delta$. We write $grade(d_j, e) = x$ ($0 \leq x \leq 1$) to denote that the assessor has decided to assign the value x to relevance of descriptor d_j for entity e. In the simplest case, a binary choice is made for x: $grade(d_j, e) = 0$ indicates that descriptor d_j does not apply to e, while $grade(d_j, e) = 1$ signifies that it does apply. If we assume $P(d_j \in e) = grade(d_j, e)$—a natural estimation—, then (2) becomes

$$P(R = 1|e, d) = 1 - \left(\prod_{j=1}^{m} (1 - grade(d_j, e) \cdot P(d|d_j)) \right). \tag{3}$$

3.4   Novelty

We now consider ranked lists of descriptors instead of single descriptors. Using (3) we can assign a score to the descriptor ranked first in the output of an entity profiling system. For descriptors returned at rank two and later, we view relevance conditioned on the descriptors ranked higher. We assume that relevance estimations have already been obtained for the first $k - 1$ descriptors in a ranked list $d_1, \ldots, d_{k-1}$ and aim to define the relevance score of descriptor d_k returned at rank k. Let the random variables associated with relevance at each rank $1, \ldots, k$ be $R_1, \ldots, R_k$. We need to estimate

$$P(R_k = 1|e, d_1, \ldots, d_k).$$

First, we estimate the degree to which support for d_k has already been provided at earlier ranks. That is, the probability that d_k contributes new information is

$$\prod_{l=1}^{k-1} (1 - P(R_l = 1|e, d_l) \cdot P(d_k|d_l)). \tag{4}$$

Here, for each descriptor d_l ranked before d_k, we determine its relevance score and use that to weight the support (if any) that d_l provides for d_k. We use (4), to replace (3) by

$$P(R_k = 1|e, d_1, \ldots, d_k) \tag{5}$$
$$= 1 - \prod_{j=1}^{m} \left(1 - grade(d_j, e) \cdot P(d_k|d_j) \cdot \prod_{l=1}^{k-1} (1 - P(R_l = 1|e, d_l) \cdot P(d_k|d_l)) \right).$$

In case the descriptors d_l (for $l \leq k - 1$) in a system-produced ranking are either non-relevant or provide no supporting evidence for the descriptor d_k returned at rank k, the terms $P(R_l|e, d_l)$ or $P(d_k|d_l)$ all attain the value 0, so that (5) reduces to (3).

3.5 Aggregating

Finally, we aggregate scores of individual descriptors into a score for ranked lists of descriptors. Discounted cumulative gain has become a standard evaluation measure when graded scores are available [11]. "Standard" notions such as gain vector, cumulative gain (CG) vector and discounted cumulative gain (DCG) vector can easily be defined, using the score produced by (5) as elements in the gain vector. I.e., the k-th element of the gain vector G is $G[k] = P(R_k = 1 | e, d_1, \ldots, d_k)$. And the cumulative gain vector is $CG[k] = \sum_{j=1}^{k} G[k]$, while the discounted cumulative gain is defined as $DCG[k] = \sum_{j=1}^{k} G[j]/(\log_2(1 + j))$. Producing the ideal cumulative gain vector (needed for computing the normalized DCG) is more complex; various approximations have been proposed. Clarke et al. [8] propose a variant α-nDCG, where α reflects the possibility of assessor error. This extension can be incorporated by modifying (3).

4 Related Work

Nugget-based evaluation methodologies have been used in a number of large-scale evaluation efforts. For the "other" questions considered at TREC 2004 and 2005, systems had to return text snippets containing important information about a topic; if the topic at hand is an entity, this boils down to entity profiling. System responses consist of passages extracted from a document collection. To evaluate responses, human assessors classify passages into essential, worthwhile (but not essential) and non-relevant [9, 15, 16]. Several authors have examined this methodology; see, e.g., [13]. A variation was considered at CLEF 2006 [12], where evaluation was again nugget-based.

Balog and de Rijke [2] considered a descriptor-based version of a specific entity profiling task, viz. expert profiling, where characteristic expertise descriptors have to be returned. Evaluation was carried out in terms of precision and recall computed against a gold standard set of descriptors for each test topic. This type of entity profiling may be viewed as a generalization of the keyphrase extraction task (assign keyphrases, typically taken from a fixed source, to a document) [10] or a variation of the query labeling task (assign labels, typically taken from a fixed knowledge source, to a query) [14].

Our evaluation framework is based on insights from [11, 12] and especially [8]. Our proposal differs from that of Clarke et al. [8] in that the items we need to assess are descriptors (not documents) and that we allow matches between system produced descriptors with gold standard descriptors to be non-exact and weighted by a semantic distance measure that fits the domain or knowledge source at hand. The theme of combining relevance assessment with semantic distance is an old one, however, going back at least 15 years [7]; to our knowledge it has so far not been applied to entity profiling.

5 Conclusion

We have developed a new scoring method for descriptor-based entity profiles that addresses many of the shortcomings of today's prevalent evaluation method. Our method allows for weighted non-exact matches between system-produced and ground-truth descriptors and it systematically rewards more highly relevant descriptors and diverse lists of descriptors. Aggregation of individual descriptor scores was done using the discounted cumulative gain measure. In future work we will explore the creation of a test collection based on the metric introduced here, taking the graded assessments in [5] as our starting point, with comparisons of different implementations of $P(d|d_j)$.

Acknowledgements. This research was supported by the European Union's ICT Policy Support Programme as part of the Competitiveness and Innovation Framework Programme, CIP ICT-PSP under grant agreement nr 250430, by the Radio Culture and Auditory Resources Infrastructure Project (LARM) funded by the Danish National Research Infrastructures Program (project nr. 09-067292), by the DuOMAn project carried out within the STEVIN programme which is funded by the Dutch and Flemish Governments under project nr STE-09-12, and by the Netherlands Organisation for Scientific Research (NWO) under project nrs 612.066.512, 612.061.814, 612.061.815, 640.004.802.

References

[1] Al-Maskari, A., Sanderson, M., Clough, P.: The relationship between IR effectivenss measures and user satisfaction. In: SIGIR 2007, pp. 773–774. ACM, New York (2007)

[2] Balog, K., de Rijke, M.: Determining expert profiles (with an application to expert finding). In: IJCAI 2007, pp. 2657–2662 (2007)

[3] Balog, K., Bogers, T., Azzopardi, L., de Rijke, M., van den Bosch, A.: Broad expertise retrieval in sparse data environments. In: SIGIR 2007, pp. 551–558. ACM, New York (2007)

[4] Balog, K., Azzopardi, L., de Rijke, M.: A language modeling framework for expert finding. Inf. Proc. & Manag. 45(1), 1–19 (2009)

[5] Balog, K., Bogers, T., van den Bosch, A., de Rijke, M.: Expertise profiling in a knowledge-intensive organization (submitted, 2010)

[6] Bogers, T., Balog, K.: UvT Expert Collection documentation. Technical report, ILK Research Group Technical Report Series no. 07-06 (July 2007)

[7] Brooks, T.A.: Topical subject expertise and the semantic distance model of relevance assessment. Journal of Documentation 51, 370–387 (1995)

[8] Clarke, C.L.A., Kolla, M., Cormack, G.V., Vechtomova, O., Ashkan, A., Büttcher, S., MacKinnon, I.: Novelty and diversity in information retrieval evaluation. In: SIGIR 2008, pp. 659–666. ACM, New York (2008)

[9] Dang, H.T., Lin, J., Kelly, D.: Overview of the TREC 2006 question answering track. In: TREC 2006. NIST (2006)

[10] Hofmann, K., Tsagkias, E., Meij, E.J., de Rijke, M.: The impact of document structure on keyphrase extraction. In: CIKM 2009, Hong Kong. ACM, New York (November 2009)

[11] Järvelin, K., Kekäläinen, J.: Cumulated gain-based evaluation of ir techniques. ACM Trans. Inf. Syst. 20(4), 422–446 (2002)

[12] Jijkoun, V., de Rijke, M.: Overview of the WiQA task at CLEF 2006. In: Peters, C., Clough, P., Gey, F.C., Karlgren, J., Magnini, B., Oard, D.W., de Rijke, M., Stempfhuber, M. (eds.) CLEF 2006. LNCS, vol. 4730, pp. 265–274. Springer, Heidelberg (2007)

[13] Lin, J., Demner-Fushman, D.: Will pyramids built of nuggets topple over. In: HLT 2006, Morristown, NJ, USA, pp. 383–390. Association for Computational Linguistics (2006)

[14] Meij, E.J., Bron, M., Huurnink, B., Hollink, L., de Rijke, M.: Learning semantic query suggestions. In: Bernstein, A., Karger, D.R., Heath, T., Feigenbaum, L., Maynard, D., Motta, E., Thirunarayan, K. (eds.) ISWC 2009. LNCS, vol. 5823, Springer, Heidelberg (2009)

[15] Voorhees, E.M.: Overview of the TREC 2004 question answering track. In: TREC 2004. NIST (2005)

[16] Voorhees, E.M., Dang, H.T.: Overview of the TREC 2005 question answering track. In: TREC 2005. NIST (2006)

Evaluating Information Extraction

Andrea Esuli and Fabrizio Sebastiani

Istituto di Scienza e Tecnologie dell'Informazione
Consiglio Nazionale delle Ricerche
56124 Pisa, Italy
{firstname.lastname}@isti.cnr.it

Abstract. The issue of how to experimentally evaluate information extraction (IE) systems has received hardly any satisfactory solution in the literature. In this paper we propose a novel evaluation model for IE and argue that, among others, it allows (i) a correct appreciation of the degree of overlap between predicted and true segments, and (ii) a fair evaluation of the ability of a system to correctly identify segment boundaries. We describe the properties of this models, also by presenting the result of a re-evaluation of the results of the CoNLL'03 and CoNLL'02 Shared Tasks on Named Entity Extraction.

1 Introduction

The issue of how to measure the effectiveness of information extraction (IE) systems has received little attention, and hardly any definitive answer, in the literature. A recent review paper on the evaluation of IE systems [1], while discussing in detail other undoubtedly important evaluation issues (such as datasets, training set / test set splits, and evaluation campaigns), devotes surprisingly little space to discussing the mathematical *measures* used in evaluating IE systems; and the same happens for a recent survey on information extraction methods and systems [2]. That the issue is far from solved is witnessed by a long discussion[1], appeared on a popular NLP-related blog, in which prominent members of the NLP community voice their discontent with the evaluation measures currently used in the IE literature, and come to the conclusion that no satisfactory measure has been found yet.

The lack of agreement on an evaluation measure for IE has several negative consequences. The first is that we do not have an agreed way to compare different IE techniques on shared benchmarks, which in itself is a hindrance to the progress of the discipline. The second is that, since IE is usually tackled via machine learning techniques, we do not have an agreed measure that learning algorithms based on explicit loss minimization can optimize. The third is that, whenever

---

[1] Christopher Manning, Hal Daume III, and others, *Doing Named Entity Recognition? Don't optimize for F_1*, http://nlpers.blogspot.com/2006/08/doing-named-entity-recognition-dont.html, accessed on July 31, 2010. The discussion is actually framed in terms of evaluating *named entity recognition* (NER), but all of it straightforwardly applies to IE tasks other than NER.

M. Agosti et al. (Eds.): CLEF 2010, LNCS 6360, pp. 100–111, 2010.
© Springer-Verlag Berlin Heidelberg 2010

we optimize the parameters of our favourite IE technique via cross-validation, we generate parameter choices that are optimal for an evaluation measure of dubious standing.

A favourite measure for evaluating IE systems is F_1 [3,4], defined as the harmonic mean of the well-known notions of *precision* (π) and *recall* (ρ):

$$F_1 = \frac{2\pi\rho}{\pi + \rho} = \frac{2\dfrac{TP}{TP+FP}\dfrac{TP}{TP+FN}}{\dfrac{TP}{TP+FP} + \dfrac{TP}{TP+FN}} = \frac{2TP}{FP + FN + 2TP} \quad (1)$$

In the IE incarnation of F_1, the symbols TP, FP, and FN stand for the numbers of true positives, false positives, and false negatives, resulting from a standard binary contingency table computed on the true and predicted "segments", where a segment is taken to be correctly recognized only when its boundaries have been exactly identified. As a result, this evaluation model is sometimes called *segmentation F-score* [5]. In this paper we argue that the segmentation F-score model has several shortcomings, and propose a new evaluation model that does not suffer from them.

The rest of the paper is organized as follows. Section 2 gives preliminary definitions. Section 3 discusses the shortcomings of the segmentation F-score model in detail, while Section 4 goes on to present our alternative model. In Section 5 we re-evaluate a number of past experiments from the literature in terms of our proposed model, and show that the two models rank competing systems in a substantively different way. Section 6 concludes by sketching avenues for future work.

2   A Formal Definition of Information Extraction

Let a text $U = \{t_1 \prec s_1 \prec \ldots \prec s_{n-1} \prec t_n\}$ consist of a sequence of *tokens* (typically: word occurrences) $t_1, \ldots, t_n$ and *separators* (typically: sequences of blanks and punctuation symbols) $s_1, \ldots, s_{n-1}$, where "$\prec$" means "precedes in the text". We use the term *textual unit* (or simply *t-unit*), with variables $u_1, u_2, \ldots$, to denote either a token or a separator. Let $C = \{c_1, \ldots, c_m\}$ be a predefined set of *tags* (aka *labels*, or *classes*), or *tagset*. Let $A = \{\sigma_{11}, \ldots, \sigma_{1k_1}, \ldots, \sigma_{m1}, \ldots, \sigma_{mk_m}\}$ be an *annotation* for U, where a *segment* σ_{ij} for U is a pair (st_{ij}, et_{ij}) composed of a *start token* $st_{ij} \in U$ and an *end token* $et_{ij} \in U$ such that $st_{ij} \preceq et_{ij}$ ("$\preceq$" obviously means "either precedes in the text or coincides with"). Here, the intended semantics is that, given segment $\sigma_{ij} = (st_{ij}, et_{ij}) \in A$, all t-units between st_{ij} and et_{ij}, extremes included, are tagged with tag $c_i{}^2$.

Given a universe of texts $\mathcal{U}$ and a universe of segments $\mathcal{A}$, we define *information extraction* (IE) as the task of estimating an unknown target function $\Phi : \mathcal{U} \times C \to \mathcal{A}$, that defines how a text $U \in \mathcal{U}$ ought to be annotated (according

---

[2] The reason why also the separators are the objects of annotation will become apparent in Section 4.

to a tagset C) by an annotation $A \in \mathcal{A}$; the result $\hat{\Phi} : \mathcal{U} \times C \to \mathcal{A}$ of this estimation is called a *tagger*. Our aim in this paper is exactly that of defining precise criteria for measuring how accurate this estimation is[3].

Given a *true annotation* $A = \Phi(U, C) = \{\sigma_{11}, \ldots, \sigma_{1k_1}, \ldots, \sigma_{m1}, \ldots, \sigma_{mk_m}\}$ and a *predicted annotation* $\hat{A} = \hat{\Phi}(U, C) = \{\hat{\sigma}_{11}, \ldots, \hat{\sigma}_{1\hat{k}_1}, \ldots, \hat{\sigma}_{m1}, \ldots, \hat{\sigma}_{m\hat{k}_m}\}$, for any $i \in \{1, \ldots, m\}$ we naturally make the general assumption that $\hat{k}_i$ may differ from k_i; that is, a tagger may in general produce, for a given tag c_i, more segments that it should, or less segments than it should.

The notion of IE we have defined allows in principle a given t-unit to be tagged by more than one tag, and might thus be dubbed *multi-tag IE*. A specific real application, also depending on the tagset considered, might have a multi-tag nature or not. For instance, in the expression "the Ronald Reagan Presidential Library" we might decree the t-units in "Ronald Reagan" to be instances of *both* the PER ("person") tag and the ORG ("organization") tag; or we might decree them to be only instances of the ORG tag. The aim of the present paper is to propose an evaluation model for IE that is intuitive and plausible irrespectively of whether the applications we are dealing with have a single-tag or a multi-tag nature. While different IE applications might want to take different stands on the single-tag vs. multi-tag issue, it is important to note that our definition above is general, since single-tag IE is just a special case of multi-tag IE. If the true set of segments is single-tag, it will be the task of the tagger to generate a single-tag prediction, and it will be the task of the evaluation model to penalize a tagger for not doing so. The multi-tag nature of our definition essentially means that, given tagset $C = \{c_1, \ldots, c_m\}$, we can split our original problem into m independent subproblems of estimating a target function $\Phi_i : \mathcal{U} \to \mathcal{A}_i$ by means of a tagger $\hat{\Phi}_i : \mathcal{U} \to \mathcal{A}_i$, for any $i \in \{1, \ldots, m\}$. Likewise, the annotations we will be concerned with from now on will actually be c_i-*annotations*, i.e., sets of c_i-*segments* of the form $A_i = \{\sigma_{i1}, \ldots, \sigma_{ik_i}\}$.

3   Problems with the Current Evaluation Model

Our proposal for evaluating IE is based on carefully distinguishing the *mathematical measure* to be adopted for evaluation from the *event space* (i.e., universe of objects) to which this measure is applied. From this point of view, we have seen in Section 1 that the standard "segmentation F-score" model of evaluating IE systems assumes F_1 as the evaluation measure and the set of segments (true or predicted) as the event space. However, this particular choice of event space is problematic. One problem is that the choice of segments as the event space makes the notion of a "true negative" too clumsy to be of any real use: a true negative should be a sequence (of any length) of tokens and separators that is neither a true nor a predicted segment, and the number of such sequences in a text of even modest length is combinatorially large, and simply too large to

---

[3] Consistently with most mathematical literature we use the caret symbol (^) to indicate estimation.

be of any use. While this does not prevent F_1 from being used as a measure, since F_1 is not a function of the number of true negatives (see Equation 1), this would not allow the use of other plausible measures of agreement between true and predicted annotation (such as e.g., Cohen's kappa [6], ROC analysis [7], or simple accuracy) that are indeed a function of the number of true negatives. A second problem is that it is not clear how partial overlap should be treated. While a true segment that perfectly coincides with a predicted segment is no doubt a true positive, when should a true segment that *partially* coincides with a predicted segment be treated as a true positive?

According to the *exact match model* (currently the most frequently used model; see e.g., [8,9,10,11,12,5]) this should never be the case. This seems too harsh a criterion: for instance, given true segment $\sigma=$ "Ronald Reagan Presidential Library" for tag ORG, a tagger that tags as ORG the segment $\hat{\sigma}=$ "Reagan Presidential Library" would receive no credit at all for this (σ would generate a false negative and $\hat{\sigma}$ would generate a false positive). Even worse, this tagger would receive even less credit than a tagger that predicts no segment overlapping with σ (this would generate a false negative but no false positive). Conversely, the (less frequently used) *overlap* model [13] returns a true positive whenever the tagger predicts a segment $\hat{\sigma}$ that overlaps even marginally with the true segment σ. This seems too lenient a criterion; in the extreme, a tagger that generates a single segment that covers the entire text U would obtain a perfect score, since every true segment overlaps with the single predicted segment. A more sophisticated variant is what we might call the *constrained overlap* model [14], in which only overlaps with at most k_1 spurious tokens and at most k_2 missing tokens are accepted as valid. This model, while less lenient, is problematic because of its dependence on parameters (k_1 and k_2), since any choice of actual values for them may be considered arbitrary. Additionally, this model does not adequately reward taggers that identify the boundaries of a segment exactly; for instance, given true segment $\sigma=$ "Ronald Reagan Presidential Library" for tag ORG, and given parameter choices $k_1 = 1$ and $k_2 = 1$, a tagger that tags as ORG the segment $\hat{\sigma}'=$ "the Ronald Reagan Presidential" is given the same credit as one that instead returns $\hat{\sigma}''=$ "Ronald Reagan Presidential Library". Similar drawbacks are presented by the *contain* model [13], which is actually a special case of the constrained overlap model in which $k_2 = 0$, and by variants of these models that have been proposed for the specific needs of biomedical NER [15]. A third problem is that, when F_1 is used as the evaluation measure and the set of segments is used as the event space, it is not clear how to deal with "tag switches", i.e., with cases in which the boundaries of a segment have been recognized (more or less exactly, according to one of the four models above) but the right tag has not (e.g., when a named entity has been correctly recognized as such but it has been incorrectly deemed as one of type PER instead of type ORG). The problems of partial overlap and tag switch may of course nastily interact, just adding to the headache.

4   The Token and Separator F_1^M Model

Essentially, the analysis of the existing IE evaluation model(s) that we have carried out in the previous section indicates that a new, improved model should (i) allow in a natural way for the notion of a "true negative", (ii) be sensitive to the *degree* of overlap between true and predicted segments, and (iii) naturally model "tag switches" and the problems arising from the presence of multiple tags in a given tagset.

4.1   The Event Space

The solution we propose is based on using *the set of all tokens and separators (i.e., the set of all t-units) as the event space*; we dub it *the token & separator model* (or *TS model*). In this solution, desideratum (i) is achieved by having true negatives consist of simple t-units, and not combinations of them; this has the advantage of being a more natural choice, and of bounding the number of true negatives by the number of t-units in the text. Desideratum (ii) is instead achieved by making the analysis more granular, and making a (true or predicted) segment contribute not one but *several* units to the contingency table, proportionally to its length. As for desideratum (iii), we will discuss how it is achieved later on in this section.

Let us assume for a moment that we stick to F_1 as the evaluation measure and that our tagset contains a single tag c_i, and let us look at the example annotated sentence of Table 1. For this example we have $F_1 = \frac{2TP}{2TP+FP+FN} = \frac{2*5}{2*5+2+1} = .769$, as deriving from the presence of 5 true positives (tokens "quick" and "brown" and the separator between them, plus tokens "lazy" and "dog"), 2 false positives (token "fox" and the separator before it) and 1 false negative (the separator between "lazy" and "dog"). The same example would have resulted in $F_1 = 0$ under the exact match model (since no segment is perfectly recognized) and $F_1 = 1$ under the overlap model (since all segments are at least partially recognized). The results according to the other two models would obviously depend on the parameter choices for k_1 and k_2.

The TS model finally makes it clear why, in the definitions of Section 2, we consider separators to be the object of tagging too: the reason is that the IE system should correctly identify segment boundaries. For instance, given the expression "Barack Obama, Hillary Clinton and Joe Biden" the perfect IE system will attribute the PER tag, among others, to the tokens "Barack", "Obama", "Hillary", "Clinton", and to the separators (in this case: blank spaces) between "Barack" and "Obama" and between "Hillary" and "Clinton", but *not* to the separator ", " between "Obama" and "Hillary". If the IE system does so, this means that it has correctly identified the boundaries of the segments "Barack Obama" and "Hillary Clinton". In the example annotated sentence of Table 1, the imperfect extraction of the second segment "lazy dog" via the two subsegments "lazy" and "dog" results in two true positives and one false negative; in this case, the system is partially penalized for having failed to recognize that

Table 1. Example sentence annotated according to a single tag c_i; A_i is the true annotation while $\hat{A}_i$ is the predicted annotation. For higher readability all tokens and separators (blanks, in this case) are numbered from 1 to 17.

A_i			c_i	c_i	c_i										c_i	c_i	c_i
	1	2	3	4	5	6	7	8	9	10	11	12	13	14	15	16	17
	The		quick		brown		fox		jumps		over		the		lazy		dog
$\hat{A}_i$			c_i	c_i	c_i	c_i	c_i								c_i		c_i

"lazy" and "dog" are not two separate segments, and that together they form a unique segment.

By moving from a model of events as segments to a more granular model of events as tokens and separators, this model thus takes in the right account the degree of overlap of true and predicted segments, and does so without resorting to numerical parameters that would require arbitrary decisions for their setting. Furthermore, by taking also separators into account, it correctly distinguishes the case of consecutive but separate segments, from the case of a single long segment consisting of their concatenation.

Note also that, in the TS model, the cardinality of the set of events (i.e., the total of the four figures in the contingency table) is fixed, since it coincides with the length $L(U) = 2n - 1$ of text U (where "length" here also takes separators into account). This is in sharp contrast with the usual model in which segments are events, since in the latter model the cardinality of the set of events depends on the prediction (i.e., a predicted annotation $\hat{A}$ that contains many segments will generate sets of events with high cardinality, and vice versa). The result of this move is that different predicted annotations are now compared with reference to the *same* contingency table, and not to different contingency tables, which is fairly foreign to the tradition of contingency-table-based evaluations.

Let us now discuss desideratum (iii) above by examining the case of a tagset $C = \{c_1, \ldots, c_m\}$ consisting of more than one tag (for the moment being let us still stick to F_1 as the evaluation measure). The TS model is naturally extended to this case by viewing the task of annotating U according to the m tags, as consisting of m essentially independent tasks. As a result, evaluation can be carried out by computing m separate contingency tables for the m individual tags $c_i \in C$, and averaging the results across the tags.

Borrowing from the tradition of information retrieval evaluation, we can either adopt *microaveraged* F_1 (denoted by F_1^{μ}) or *macroaveraged* F_1 (F_1^M). F_1^{μ} is obtained by (i) computing the category-specific values TP_i, FP_i and FN_i, (ii) obtaining TP as the sum of the TP_i's (same for FP and FN), and then (iii) applying Equation (1). F_1^M is instead obtained by first computing the category-specific F_1 values and then averaging them across the c_i's.

It is well-known (see e.g., [16]) that, for the same annotation, F_1^{μ} and F_1^M may yield very different results. In fact, F_1^{μ} tends to be heavily influenced by the results obtained for the more frequent tags, since for these tags TN, FN and FP (the only arguments of the F_1 function) tend to be higher than for the infrequent tags. F_1^M has instead a more "democratic" character, since it gives the same importance to every tag in the tagset. As a result, it tends to return lower values than the (somehow overoptimistic) ones returned by F_1^μ, and to reward the systems that behave well also on the more infrequent tags. Because of this important property, we propose the adoption of *macroaveraging* as the default way of averaging results across tags.

A potential criticism of the fact that tagging under tagset $C = \{c_1, \ldots, c_m\}$ is evaluated as consisting of m independent tasks, is that certain tag switches may result in too severe a penalty. For instance, a system that correctly identifies the boundaries of segment "San Diego" but incorrectly tags it as PER instead of LOC is assigned three false negatives (for failing to recognize the LOC character of the segment) *and* three false positives (for incorrectly deeming the segment an instance of PER). We feel that this is actually not too severe a penalty in the general case in which the two involved tags are not known to be close in meaning. For instance, in an opinion extraction task (see e.g., [17]), the AGENT tag (that denotes either the source or the target agent of a "private state") and the DIRECT-SUBJECTIVE tag (that denotes either the explicit mention of a private state or a speech event expressing a private state) denote two concepts very distant in meaning, so distant that it seems reasonable to evaluate a tag switch between them as involving *both* false positives and false negatives. Conversely, in a task such as NER in which the different tags (PER, LOC, ORG, MISC) are close in meaning, the tags may be viewed as subtags of a common supertag ("ENTITY"). If desired, a more lenient evaluation may be performed by also evaluating ENTITY as a tag in its own. At this less granular level, correctly identifying the boundaries of a LOC segment but mistagging it as PER, would only give rise to true positives; this would provide, when desired, a coarser level of analysis that is more lenient towards tag switches between semantically related tags.

4.2   The Evaluation Measure

Concerning the evaluation measure to adopt, it is interesting to see that, in combination with the TS model, the problems that had plagued F_1 (and that had prompted "Don't optimize for F_1!" recommendations – see Footnote 1) disappear, which makes F_1 a plausible evaluation measure for IE. Concerning this, an interesting property of F_1 is that it does not depend on true negatives, which are going to be in very high numbers in many IE applications such as NER; in other words, F_1 is inherently robust to the typical high imbalance between the positive and the negative examples of a tag. A second interesting property of F_1 is that it does not encourage a tagger to either undertag or overtag, since the trivial rejector (i.e., the tagger that does not tag any t-unit) has an F_1 score of 0, and the trivial acceptor (i.e., the tagger that tags all t-units) has an F_1 score equal to the fraction of true tagged t-units, which is usually very low. A third useful property of F_1 is that its more general form ($F_\beta = \frac{(\beta^2+1)\pi\rho}{\beta^2\pi+\rho} = \frac{(\beta^2+1)TP}{TP+FP+\beta^2(TP+FN)}$ – see e.g., [3]) also allows, if needed, a higher penalty to be placed on overtagging than undertagging (this is accomplished by picking a value of β in $[0,1)$, with lower values placing heavier penalties) or viceversa (β in $(1,+\infty)$, with higher values placing heavier penalties). Last, it should be mentioned that learning algorithms for IE that are capable of internally optimizing for F_1 are available (in both the support vector machines camp – see [18] – and the conditional random fields camp – see [5]), thus making it possible to generate taggers that are accurate at maximizing the two factors that our TS model rewards, i.e., (i) the degree of overlap between true and predicted segments, and (ii) the ability to correctly identify segment boundaries.

Given the fact that we advocate using (a) the set of tokens and separators as the set of events, (b) F_1 as the evaluation function, and (c) macroaveraging as the method for averaging results across tags, we will henceforth refer to our proposed model as *the token & separator F_1^M model* (or *TS-F_1^M model*).

5   Experiments

In order to provide an indication of the impact that our proposed model may have on a concrete evaluation, we have re-evaluated according to the TS-F_1^M model the submissions to the CoNLL'03 [12][4] and CoNLL'02 [11][5] Named Entity Extraction Shared Tasks. The CoNLL'03 NER Shared Task attracted 16 participants, and consisted of two subtasks, one on English and the other on German NER. The CoNLL'02 NER Shared Task attracted instead 12 participants, who competed on both Spanish and Dutch NER. We here deal only with the 2003 English and German data and with the 2002 Spanish data; we could not re-evaluate the 2002 Dutch data since the original files are no longer available due to copyright problems[6].

The 1st row of Table 2 presents the way the 16 participants on 2003 English data are ranked according to the segmentation F-score ("segment-based, exact-match F_1^μ model", in our terminology) officially adopted in the shared task, while the 2nd row reports the same for the TS-F_1^M model. Although the two rankings are not too dissimilar (e.g., the first 4 positions are the same), there are a few relevant differences. The participant that originally placed 11th in CoNLL'03 is ranked in 5th position by our evaluation model, jumping no less than 6 positions up in the ranking. This indicates that the algorithm of the 11th participant was perhaps suboptimal at producing exact matches (it indeed generated 2.3% fewer exact matches than the 5th participant) but often generated predicted segments closely corresponding to the true segments (e.g., it indeed generated 157.6% more "close matches" – i.e., accurate modulo a single token – than the 5th participant, and totally missed 6.9% fewer segments than the 5th participant). Conversely, our evaluation method demotes by 3 places each the participants that originally placed 5th, 7th and 12th. Several other participants are promoted or demoted

---

[4] http://www.cnts.ua.ac.be/conll2003/ner/
[5] http://www.cnts.ua.ac.be/conll2002/ner/
[6] Erik Tjong Kim Sang, Personal communication, 25 Feb 2010.

Table 2. Rankings of the CoNLL'03 (English and German) and CoNLL'02 (Spanish) Shared Tasks participants according to the segment (Seg), token (T), and token & separator (TS) event spaces and to measures F_1^μ and F_1^M. The value in each cell represents the original rank the system obtained in the CoNLL'03 / CoNLL'02 evaluations, which use a segment-based F_1^μ exact-match model (1st, 4th, and 7th rows).

ENGLISH Seg-F_1^μ	1	2	3	4	5	6	7	8	9	10	11	12	13	14	15	16
	.888	.883	.861	.855	.850	.849	.847	.843	.840	.839	.825	.817	.798	.782	.770	.602
TS-F_1^M	1	2	3	4	11	8	6	5	10	7	9	14	15	13	12	16
	.875	.874	.857	.853	.848	.845	.842	.840	.835	.833	.819	.817	.813	.809	.808	.671
T-F_1^M	1	2	3	4	11	6	8	5	7	10	9	14	15	12	13	16
	.885	.880	.865	.863	.857	.855	.853	.848	.847	.846	.833	.824	.822	.821	.815	.699
GERMAN Seg F_1^μ	1	2	3	4	5	6	7	8	9	10	11	12	13	14	15	16
	.724	.719	.713	.700	.692	.689	.684	.681	.678	.665	.663	.657	.630	.573	.544	.477
TS-F_1^M	1	9	3	2	4	7	6	5	8	11	10	13	12	14	15	16
	.719	.708	.706	.702	.695	.691	.690	.679	.674	.650	.645	.642	.641	.616	.569	.471
T-F_1^M	1	9	3	2	4	7	6	5	8	11	10	13	12	14	15	16
	.726	.714	.713	.709	.701	.699	.697	.685	.680	.671	.652	.647	.644	.621	.582	.478
SPANISH Seg F_1^μ	1	2	3	4	5	6	7	8	9	10	11	12				
	.814	.791	.771	.766	.758	.758	.739	.739	.737	.715	.637	.610				
TS-F_1^M	1	2	4	5	6	7	10	9	8	3	12	11				
	.821	.799	.769	.746	.746	.740	.734	.729	.724	.710	.677	.636				
T-F_1^M	1	2	4	5	6	7	10	9	8	3	12	11				
	.823	.804	.775	.752	.752	.749	.741	.737	.732	.721	.681	.648				

by 2 places. The results are not much different in the 2003 German and 2002 Spanish data. In the 2003 German task we even have a participant that gains 8 positions (from 9th to 2nd place), while in the 2002 Spanish task one participant is downgraded from 3rd to 10th place (3rd from last!). These potentially very large differences clearly indicate that taking a clear stand between the two models is essential. In order to check the level of correlation of the two models we have also computed (see Table 3) the Spearman's rank correlation

$$R(\eta', \eta'') = 1 - \frac{6\sum_{k=1}^{p}(\eta'(\hat{\Phi}_k) - \eta''(\hat{\Phi}_k))^2}{p(p^2 - 1)}$$

between the rankings η' and η'' generated by the two models, where p is the number of ranked participants, and $\eta(\hat{\Phi}_k)$ denotes the rank position of system $\hat{\Phi}_k$ in ranking η. We can see from Table 3 (whose values are obtained by averaging across the $R(\eta', \eta'')$ values obtained in the English, German and Spanish tasks) that the rankings produced by the two models are fairly correlated ($R = .832$) but not *highly* so, confirming that taking a stand between the two is indeed important.

A potential criticism to using the set of all t-units as the event space (instead of, say, the set of all tokens) is that separators are given the same importance as tokens, which might seem excessive. For instance, the imperfect identification of

Table 3. Spearman's rank correlation (averaged across the English, German, and Spanish tasks) $R(\eta', \eta'')$ between the results produced by the three evaluation models discussed in this section

	Seg F_1^μ	TS-F_1^M	T-F_1^M
Seg F_1^μ	1.0	.832	.832
TS-F_1^M	.832	1.0	.990
T-F_1^M	.832	.990	1.0

the true segment "Barack Obama" via the predicted segment "Obama" results in one true positive and not one but *two* false negatives, which might be deemed too harsh a penalty. A potential solution to this problem consists in *weighting* tokens and separators differently, since F_1 can handle "weighted events" seamlessly. For instance, if we weigh separators half as much as tokens, a correctly tagged separator will count as "half a true positive"; accordingly, the example annotated sentence of Table 1 would obtain $F_1 = \frac{2*4.5}{2*4.5+1.5+0.5} = .818$. A similar solution could be adopted if different types of tokens are deemed to have different importance; for instance, heads might be weighted higher than modifiers in some applications, and last names might be weighted higher than first names when extracting person names.

Anyhow, in order to assess whether giving separators the same importance as tokens indeed constitutes a problem, we have re-evaluated the CoNLL'03 and CoNLL'02 results also according to a "token-only" F_1^M model (hereafter dubbed *T-F_1^M model*), i.e., a model which differs from our proposed model in that separators are not part of the event space, and are thus not the object of evaluation. The results are reported in the 3rd, 6th and 9th rows of Table 2. For the English data, we can see that the rankings are fairly similar, with only a few systems swapping places with the system next in the ranking (this happens for the systems placed 6th, 9th, and 14th in the TS-F_1^M ranking). For the German and Spanish data, the rankings are identical to the ones of the TS-F_1^M ranking. As a result, the Spearman's rank correlation R between the two rankings is very high ($R = .990$). All this indicates that the TS-F_1^M model does not place excessive emphasis on separators, which is good news.

Additionally, we should consider that separators tend to have even more negligible effects in IE tasks characterized by segments longer than the ones to be found in the CoNLL NER tasks. To see this, assume that, given a true segment σ containing n tokens, a tagger $\hat{\Phi}$ correctly recognizes only its subsegment containing the first $\frac{1}{2}n$ tokens. If $n = 2$, $\hat{\Phi}$ will obtain precision values of $\pi = \frac{1}{3}$ or $\pi = \frac{1}{2}$ (a very substantive difference) according to whether separators are considered or not in the evaluation. If $n = 100$, instead, $\hat{\Phi}$ will obtain precision values of $\pi = \frac{99}{199}$ or $\pi = \frac{50}{100}$, whose difference is almost negligible. Similar considerations hold for recall.

All in all, given that the difference between the rankings produced by the TS-F_1^M model and by the T-F_1^M model is small, and given that the former offers better theoretical guarantees than its token-only counterpart (since it guarantees that the correct identification of segment boundaries is properly rewarded), we think that the former should be preferred to the latter.

A scorer that evaluates a text annotated in the common IOB2 format according to *both* the segmentation F-score and the TS-F_1^M model can be downloaded at http://patty.isti.cnr.it/~esuli/IEevaluation/

6   Conclusion

We have argued that, in order to overcome the shortcomings of the standard "segmentation F-score" evaluation model for IE, the choices of event space and evaluation measure should be considered as two separate issues. For the former, we have proposed using as the event space the set of all tokens and separators. We have shown that this (i) allows a correct appreciation of the degree of overlap between predicted and true segments, (ii) allows a fair evaluation of the ability of a system to correctly identify segment boundaries, (iii) has the consequence that the notion of a "true negative" is clearly defined, and (iv) allows the comparative evaluation of different IE systems to be carried out on the same contingency table. We have also argued that "tag switches" do not pose evaluation problems once different evaluations are carried out independently for different tags and then averaged. As for the evaluation measure, we have argued that, although there is nothing wrong with sticking to the standard F_1 measure, its macroaveraged version (F_1^M) is somehow more desirable, since it rewards systems that perform well across the entire tagset.

Finally, we should note that the notion of IE we have defined also allows a given t-unit to belong to more than one segment *for the same tag* c_i (we might thus dub this *multi-instance IE*). While this situation never occurs in simple applications of IE such as NER, there exist instances of IE in which this is the case. For example, in the tagset for opinion extraction defined in [17], it does happen that the same t-unit may belong to several segments for the same tag; e.g., in sentence "John wrote me that Mary said I love pizza", the segment "I love pizza" belongs to *two* overlapping segments of the INSIDE tag. Both the segmentation F-score and the evaluation model we have presented in this paper can only handle the single-instance IE case; we leave the issue of how to best evaluate multi-instance IE to further research.

References

1. Lavelli, A., Califf, M.E., Ciravegna, F., Freitag, D., Giuliano, C., Kushmerick, N., Romano, L., Ireson, N.: Evaluation of machine learning-based information extraction algorithms: Criticisms and recommendations. Language Resources and Evaluation 42(4), 361–393 (2008)
2. Sarawagi, S.: Information extraction. Foundations and Trends in Databases 1(3), 261–377 (2008)

3. Lewis, D.D.: Evaluating and optmizing autonomous text classification systems. In: Proceedings of the 18th ACM International Conference on Research and Development in Information Retrieval (SIGIR 1995), Seattle, US, pp. 246–254 (1995)

4. van Rijsbergen, C.J.: Foundations of evaluation. Journal of Documentation 30(4), 365–373 (1974)

5. Suzuki, J., McDermott, E., Isozaki, H.: Training conditional random fields with multivariate evaluation measures. In: Proceedings of the 21st International Conference on Computational Linguistics and 44th Annual Meeting of the ACL (ACL/COLING 2006), Sydney, AU, pp. 217–224 (2006)

6. Cohen, J.: A coefficient of agreement for nominal scales. Educational and Psychological Measurement 20(1), 37–46 (1960)

7. Fawcett, T.: An introduction to ROC analysis. Pattern Recognition Letters 27, 861–874 (2006)

8. Freitag, D., Kushmerick, N.: Boosted wrapper induction. In: Proceedings of the 17th Conference of the American Association for Artificial Intelligence (AAAI 2000), Austin, US, pp. 577–583 (2000)

9. Freitag, D.: Machine learning for information extraction in informal domains. Machine Learning 39, 169–202 (2000)

10. Krishnan, V., Manning, C.D.: An effective two-stage model for exploiting non-local dependencies in named entity recognition. In: Proceedings of the 21st International Conference on Computational Linguistics and 44th Annual Meeting of the Association for Computational Linguistics (COLING/ACL 2006), Sydney, AU, pp. 1121–1128 (2006)

11. Tjong Kim Sang, E.F.: Introduction to the CoNLL-2002 shared task: Language-independent named entity recognition. In: Proceedings of the 6th Conference on Natural Language Learning (CONLL 2002), Taipei, TW, 155–158 (2002)

12. Tjong Kim Sang, E.F., De Meulder, F.: Introduction to the CoNLL-2003 shared task: Language-independent named entity recognition. In: Proceedings of the 7th Conference on Natural Language Learning (CONLL 2003), Edmonton, CA, pp. 142–147 (2003)

13. Freitag, D.: Using grammatical inference to improve precision in information extraction. In: Proceedings of the ICML 1997 Workshop on Automata Induction, Grammatical Inference, and Language Acquisition, Nashville, US (1997)

14. De Sitter, A., Daelemans, W.: Information extraction via double classification. In: Proceedings of the ECML/PKDD 2003 Workshop on Adaptive Text Extraction and Mining, Cavtat-Dubrovnik, KR, pp. 66–73 (2003)

15. Tsai, R.T.H., Wu, S.H., Chou, W.C., Lin, Y.C., He, D., Hsiang, J., Sung, T.Y., Hsu, W.L.: Various criteria in the evaluation of biomedical named entity recognition. BMC Bioinformatics 7(92) (2006)

16. Sebastiani, F.: Machine learning in automated text categorization. ACM Computing Surveys 34(1), 1–47 (2002)

17. Wiebe, J., Wilson, T., Cardie, C.: Annotating expressions of opinions and emotions in language. Language Resources and Evaluation 39(2/3), 165–210 (2005)

18. Joachims, T.: A support vector method for multivariate performance measures. In: Proceedings of the 22nd International Conference on Machine Learning (ICML 2005), Bonn, DE, 377–384 (2005)

Tie-Breaking Bias: Effect of an Uncontrolled Parameter on Information Retrieval Evaluation*

Guillaume Cabanac, Gilles Hubert,
Mohand Boughanem, and Claude Chrisment

Université de Toulouse — IRIT UMR 5505 CNRS
118 route de Narbonne, F-31062 Toulouse cedex 9
{cabanac,hubert,boughanem,chrisment}@irit.fr

Abstract. We consider Information Retrieval evaluation, especially at TREC with the trec_eval program. It appears that systems obtain scores regarding not only the relevance of retrieved documents, but also according to document names in case of ties (i.e., when they are retrieved with the same score). We consider this tie-breaking strategy as an uncontrolled parameter influencing measure scores, and argue the case for fairer tie-breaking strategies. A study of 22 TREC editions reveals significant differences between the Conventional unfair TREC's strategy and the fairer strategies we propose. This experimental result advocates using these fairer strategies when conducting evaluations.

1   Introduction

Information Retrieval (IR) is a field with a long tradition of experimentation dating back from the 1960s [1]. The IR community notably benefited from TREC evaluation campaigns and workshops. Since 1992, these have been offering researchers the opportunity to measure system effectiveness and discuss the underlying theoretical aspects [2,3]. At TREC, evaluation results of IR systems (IRSs), a.k.a. search engines, are computed by the trec_eval [4] program. Many subsequent IR evaluation initiatives also rely on trec_eval, such as tasks of NTCIR [5], CLEF [6], and ImageCLEF [7].

As a general rule when conducting a scientific experiment, one should identify all the parameters at stake and control all but one to be able to test its effect on the measured artifact. Controlling parameters is a key concern since conclusions may be biased when two or more parameters vary at the same time during the experiment. Following on from studies on IR evaluation methodology such as Voorhees's [8] and Zobel's [9] we identified an uncontrolled parameter in TREC through trec_eval: evaluation results not only depend on retrieved documents, but also on how they were named in case of ties (i.e., *ex aequo* documents). This is a major issue since 'lucky' ('unlucky') IRSs can get better (worse) results than they would deserve in an unbiased evaluation.

---

* This work was partly realized as part of the Quaero Programme, funded by OSEO, French State agency for innovation.

M. Agosti et al. (Eds.): CLEF 2010, LNCS 6360, pp. 112–123, 2010.
© Springer-Verlag Berlin Heidelberg 2010

This paper is organized as follows. In Sect. 2, we present how IRSs are commonly evaluated according to the Cranfield paradigm. Then, we detail in Sect. 3 the issue we identified in TREC methodology, that we call the 'tie-breaking bias.' In Sect. 4, we propose alternative reordering strategies for canceling out the effect of the considered uncontrolled parameter. A comparison of the current TREC strategy with our proposal is detailed in Sect. 5. Our findings and limitations of our analyses are discussed in Sect. 6. Finally, related works are reviewed in Sect. 7 before concluding the paper and giving insights into research directions.

2   Tie-Breaking Prior to Evaluating IRSs Effectiveness

This section introduces the concepts considered throughout the paper. Our brief description may be complemented by [10], which details IR evaluation at TREC and its realization with the trec_eval program. At TREC, at least one *track* a year is proposed. A track is comprised of 50+ *topics*; each one is identified with a qid. Participants in a track contribute at least one *run* file. Among the fields of this file, trec_eval only considers the following: qid, the document identifier docno, and the similarity sim of docno regarding qid, as provided by the IRS. In addition, the *query relevance judgments* file (i.e., *qrels*) results from manual assessment. The trec_eval program only considers the 3 following fields from it: qid, docno, and rel, which represents the relevance of docno regarding qid. The rel $\in [1, 127]$ value is assigned to relevant documents. Other values (e.g., rel $= 0$) represent non-relevant documents. Prior to computing effectiveness measures, trec_eval pre-processes the run file. Since it ignores its rank field, documents are reordered as follows: "internally ranks are assigned by sorting by the sim field with ties broken deterministicly (using docno)" [4]. Buckley and Voorhees comment this rationale and underline the importance of tie-breaking:

> "For TREC-1 ... Each document was also assigned a rank by the system, but this rank was deliberately ignored by trec_eval. Instead, trec_eval produced its own ranking of the top two hundred documents[1] based on the RSV [sim] values to ensure consistent system-independent tie breaking among documents that a system considered equally likely to be relevant (the ordering of documents with tied RSV values was arbitrary yet consistent across runs). Breaking ties in an equitable fashion was an important feature at the time since many systems had large number of ties—Boolean and coordination-level retrieval models could produce hundreds of documents with the same RSV." [10, p. 55]

Finally, trec_eval uses qrels and the reordered run to compute several effectiveness measures. In the remainder of this paper, let us consider a system s, a topic t, and a document d of the run, and the following measures: Reciprocal Rank $RR(s, t)$ of top relevant document, Precision at d cutoff $P(s, t, d)$, Average Precision $AP(s, t)$, and Mean Average Precision $MAP(s)$. Due to space limitation, we do not elaborate on these measures and refer the reader to [11, ch. 8] for a comprehensive definition.

---

[1] Since TREC-2, the top 1,000 documents is kept [10, p. 58].

The next section presents an issue related to the way document ties are broken at TREC. We argue that this issue makes the current tie-breaking strategy an uncontrolled parameter in IR experiments.

3   On How the Tie-Breaking Bias Influences IR Evaluation

Let us consider, in Fig. 1(a), a sample of a run concerning the top 3 documents retrieved by an IRS for a given topic t (qid = 3). Suppose that 5 documents are relevant documents for t in the collection, including **WSJ5** (in bold). Since trec_eval ignores ranks it reorders the run by ascending qid, descending sim, and descending docno for tie-breaking purpose. The resulting document list is presented in Fig. 1(b) where the relevant **WSJ5** document is assigned rank #1. Notice that reciprocal rank is $RR(s,t) = 1$, precision at **WSJ5** is $P(s,t,$**WSJ5**$) = 1$ and $AP(s,t) = 1/5$. Now, without making any changes to the document contents, which still remain relevant for topic t, suppose that **WSJ5** had been named **AP8** instead. So, relevant document **AP8** is initially ranked #2 (i.e., the same position as **WSJ5**), as shown in Fig. 1(c). Then, due to the reordering process, LA12 remains ranked #1 by descending docno, remaining above **AP8**. Notice that reciprocal rank and average precision have been halved.

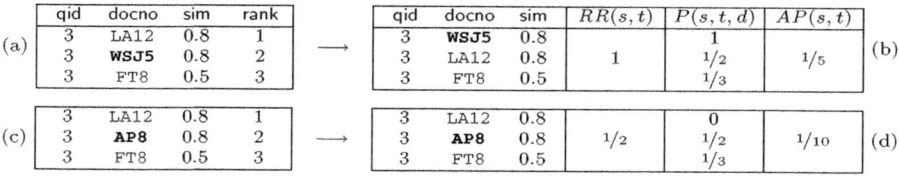

Fig. 1. Effect of document naming on reordered run and measure values

This minimal example illustrates the issue addressed in the paper: IRS scores depend not only on their ability to retrieve relevant documents, but also on document names in case of ties. Relying on docno field for breaking ties here implies that the *Wall Street Journal* collection (WSJ⋆ documents) is more relevant than the *Associated Press* collection (AP⋆ documents) for whatever the topic, which is definitely wrong. This rationale introduces an uncontrolled parameter in the evaluation regarding all rank-based measures, skewing comparisons unfairly. Let us justify our statement by considering the example of AP:

1) Tie-breaking effect on *inter-system* comparison, where $AP(s_1,t)$ of system s_1 and $AP(s_2,t)$ of system s_2 are considered for a given topic t. This comparison is unfair since AP values can be different although both the systems returned the same result $[R_{0.8}, N_{0.8}, N_{0.5}]$ where R_x is a relevant document (N_x is a nonrelevant document) retrieved with sim $= x$. This is the case when we associate the run in Fig. 1(a) with s_1, and the run in Fig. 1(c) with s_2. Indeed $AP(s_1,t) = 1/1 \cdot 1/5$ whereas $AP(s_2,t) = 1/2 \cdot 1/5 = 1/10$, thus showing a 200% difference.

2) Tie-breaking effect on *inter-topic* comparison, where we consider $AP(s, t_1)$ and $AP(s, t_2)$ of a single system for two topics t_1 and t_2. Such a comparison is made in TREC's *robust* [12] track for characterizing easy and difficult information needs. It is unfair since TREC reordering process may have benefited system s for t_1 (by re-ranking relevant tied documents upwards in the list) while having hindered it for t_2 (by re-ranking relevant tied documents downwards in the list). As a result, the IRS designers may conduct failure analysis to figure out why their system poorly performed on some topics. Poor results, however, may only come from misfortune when relevant documents are reorganized downwards in the result list only because of their names. Imagining that every relevant document comes from the AP collection, they will be penalized since they will be re-ranked at the bottom of the tied group when reordering by decreasing docno.

Breaking ties as currently proposed at TREC introduces an uncontrolled parameter affecting IR evaluation results. In order to avoid this tie-breaking issue, the next section introduces our proposal: alternative reordering strategies.

4   Realistic and Optimistic Tie-Breaking Strategies

The current tie-breaking strategy (qid asc, sim desc, docno desc) introduces an uncontrolled parameter, as it relies on the docno field for reordering documents with the same sim value. Another strategy would be to randomize tied documents; this is not suitable as evaluations would be unfair and not reproducible (non deterministic). However, evaluations must measure how well a contribution performed, not how well chance benefited an IRS. Alternatively, relying on the initial ranks (from run) implies the same issue: IRS designers may have untied their run by assigning random ranks, as they were not able to compute a discriminative sim for those documents. As a result, random-based and initial rank-based approaches do not solve the tie-breaking issue.

In this section, we propose two tie-breaking strategies that are not subject to the bias presented in this paper. Figure 2 shows merged runs and qrels, as well as the result of current TREC Conventional strategy for reordering ties and the two strategies that we propose:

1. *Realistic reordering* stipulates that tied nonrelevant documents should come above relevant documents in the ranked list because the IRS was not able to differentiate between them. The reordering expression meeting this requirement is "qid asc, sim desc, rel asc, docno desc."

$$\text{Example: } [R_x, N_x, R_x] \xrightarrow[\text{qid asc, sim desc, rel asc, docno desc}]{\textit{Realistic} \text{ reordering}} [N_x, R_x, R_x].$$

2. *Optimistic reordering* stipulates that tied relevant documents should come above nonrelevant documents in the ranked list because the IRS may present them together, within clusters for instance. The reordering expression meeting this requirement is "qid asc, sim desc, rel desc, docno desc."

$$\text{Example: } [R_x, N_x, R_x] \xrightarrow[\text{qid asc, sim desc, rel desc, docno desc}]{\textit{Optimistic} \text{ reordering}} [R_x, R_x, N_x].$$

Regarding the selected reordering strategy (Realistic, Conventional or Optimistic) the value of a measure can differ. Notice that Optimistic reordering is fair but game-able, which is bad: a run comprised of ties only would be evaluated just like if it ranked all relevant documents at the top. As a result, we recommend the use of Realistic strategy for conducting fair evaluations. In the remainder of the paper, 'M_S' denotes measure M with reordering strategy $S \in \{R, C, O\}$. Notice the total order $M_R \leqslant M_C \leqslant M_O$ between measure values.

qid	docno	sim	rank	rel
8	**CT5**	0.9	1	1
8	AP5	0.7	2	0
8	WSJ9	0.7	3	0
8	**AP8**	0.7	4	1
8	FT12	0.6	5	0

qid	docno	sim	rel
8	**CT5**	0.9	1
8	WSJ9	0.7	0
8	AP5	0.7	0
8	**AP8**	0.7	1
8	FT12	0.6	0

(b) Realistic reordering

qid	docno	sim	rel
8	**CT5**	0.9	1
8	WSJ9	0.7	0
8	**AP8**	0.7	1
8	AP5	0.7	0
8	FT12	0.6	0

(c) Conventional reordering

qid	docno	sim	rel
8	**CT5**	0.9	1
8	**AP8**	0.7	1
8	WSJ9	0.7	0
8	AP5	0.7	0
8	FT12	0.6	0

(d) Optimistic reordering

Fig. 2. Realistic, Conventional, and Optimistic reordering strategies for a run

We demonstrated in this section how an uncontrolled parameter (i.e., document naming) affects IRSs scores. In order to foster fairer evaluation, we proposed alternative Realistic and Optimistic reordering strategies. In the next section, we conduct an analysis of past TREC datasets to measure the effect of the chosen reordering strategy on evaluation results.

5   Effect of the Tie-Breaking Bias on IR Evaluation

We studied the effect of the tie-breaking bias on the results of 4 TREC tracks: *ad hoc* (1993–1999), *routing* (1993–1997), *filtering* (limited to its *routing* subtask, 1998–2002, as other subtasks feature binary sim values, making them inappropriate for our study), and *web* (2000–2004). The corresponding 22 editions comprise 1,360 *runs* altogether, whose average length is 50,196 lines. This represents 3 Gb of raw data retrieved from trec.nist.org and analyzed as follows. In Sect. 5.1, we evaluate to what extent runs are concerned with the uncontrolled parameter issue by assessing the proportion of document ties within runs. Then, in Sect. 5.2, we report the differences between scores obtained with the proposed fair reordering strategies *vs* the Conventional strategy promoted at TREC.

5.1   Proportion of Document Ties as Observed in 22 Trec Editions

In the remainder of the paper, we call a *result-list* the sample of a run concerning a specific topic qid submitted in a given *year*, and denote it $\mathrm{runid}_{\mathsf{qid}}^{year}$. Since differences in scores arise when a result-list contains tied documents, this section assesses how often such a phenomenon happened in the considered TREC dataset. Table 1 shows statistics related to each track: the considered editions (Year) and number of submitted runs (detailed for each year, and overall). Two other indicators are provided, regarding *the percentage of ties in result-lists, and ☆the average number of tied documents when grouped by equal similarity (sim). Statistics related to minimum (Min), average (Avg), maximum (Max) and standard deviation (SD) are also reported. For instance, the result-list featured in Fig. 2 contains *3/5 = 60% of tied documents, and ☆presents an average of $(1+3+1)/3 = 1.7$ tied documents per sim.

Table 1. Proportion of document ties as observed in the runs of 4 TREC tracks

Track	Year	# of runs	*Tied docs in a result-list (%)				☆Avg # of tied docs per sim			
			Min	Avg	Max	SD	Min	Avg	Max	SD
ad hoc	1993	36	0.0	30.3	100.0	36.0	2.2	4.4	28.0	4.2
	1994	40	0.0	28.4	100.0	35.9	1.9	9.5	37.3	11.2
	1995	39	0.0	29.2	99.9	32.8	1.0	2.8	26.2	4.2
	1996	82	0.0	24.1	100.0	32.3	2.0	4.1	35.1	4.7
	1997	79	0.0	24.7	100.0	34.7	1.8	4.5	25.8	5.1
	1998	103	0.0	19.0	100.0	27.4	1.0	2.5	33.8	4.4
	1999	129	0.0	15.6	100.0	24.6	1.5	3.7	22.9	4.4
	Avg over 508 *runs* →		0.0	24.5	100.0	32.0	1.6	4.5	29.9	5.5
filtering	1998	47	0.0	26.8	100.0	40.8	41.0	42.0	51.8	2.2
	1999	55	0.0	7.5	100.0	23.8	2.1	2.1	2.7	0.1
	2000	53	0.0	21.1	100.0	38.1	15.3	22.3	37.1	10.0
	2001	18	0.0	25.6	100.0	30.3	19.8	33.3	69.6	17.0
	2002	17	0.0	34.6	100.0	37.2	2.5	23.3	97.9	33.2
	Avg over 190 *runs* →		0.0	23.1	100.0	34.0	16.1	24.6	51.8	12.5
routing	1993	32	0.0	32.9	100.0	39.9	1.1	4.1	38.2	6.0
	1994	34	0.0	31.0	100.0	37.6	2.3	5.5	30.9	5.9
	1995	27	0.0	24.9	99.2	27.4	1.0	1.5	14.7	1.4
	1996	26	0.0	21.3	100.0	24.5	1.4	7.2	40.0	10.7
	1997	34	0.0	27.4	100.0	33.7	6.7	13.0	54.3	10.9
	Avg over 153 *runs* →		0.0	27.5	99.8	32.6	2.5	6.3	35.6	7.0
web	2000	104	0.0	29.3	100.0	34.3	2.9	9.3	79.6	16.6
	2001	96	0.0	32.0	100.0	31.9	25.8	27.8	63.8	5.7
	2002	71	0.0	25.8	100.0	33.5	1.0	3.6	44.7	6.3
	2003	164	0.0	18.8	100.0	27.8	1.4	2.3	12.0	1.8
	2004	74	0.0	24.9	100.0	34.4	1.5	4.3	39.6	6.2
	Avg over 509 *runs* →		0.0	26.2	100.0	32.4	6.5	9.5	47.9	7.3
	Total avg over 1,360 *runs* →		0.0	25.2	100.0	32.7	6.2	10.6	40.3	7.8

Overall, IRSs participating in early TREC *ad hoc* editions contributed more result-lists with tied documents than later on. This is in line with Buckley and Voorhees's observation [10, p. 55] quoted in Sect. 2.

Averaging over each track, 25.2% of a result-list is comprised of tied documents. This proportion is highly variable, as highlighted by an average 32.7% standard deviation (strikingly similar for each track). Moreover, each year featured result-lists with no ties at all (i.e., Min* = 0.0). It also happened that some result-lists consisted of tied documents only (1,338 result-lists over the 4 tracks). The latter case may be illustrated at TREC *ad hoc* by ibmge2 $_{291}^{1996}$

as an example of non-discrimination: all retrieved documents share the same sim $= -126.000000$ score. Those result-lists are most likely to obtain completely different results according to the applied tie-breaking strategy.

Regarding a run, when we consider the retrieved documents grouped by sim, we notice a great variability. Some result-lists have no ties ($\text{Min}^{\star} = 1.0$, which corresponds to $\text{Min}^{\star} = 0.0$) while others have on average up to 97.9 documents with the same sim value. The average group size of 10.6 documents implies that a document ranked at position $r + 10$ with Realistic strategy can be re-ranked rth with another strategy if lucky enough. Generalizing this observation, the larger the tied document group is, the larger the unfair position gain or loss will be.

This section showed that result-lists are likely to contain several tied documents. Thus, the uncontrolled parameter that we identified may affect IR evaluation. This hypothesis is tested in the next section.

5.2   Result Differences regarding the Adopted Reordering Strategy

We emphasize comparisons between M_R and M_C (i.e., Realistic vs Conventional) to show how luck increased the M_R score deserved in a fair and unbiased setting. Notice however that larger differences will be observed when comparing M_R and M_O as the tested measures are totally ordered: $M_R \leqslant M_C \leqslant M_O$, cf. Sect. 4. For each measure, we first present systems that most benefited from the uncontrolled parameter by showing the top 3 differences between unfair M_C and fair M_R. Then, we generalize these findings by reporting statistical significance of $M_C - M_R$ for each track as a whole (i.e., considering every contributed runs). Significance p-values result from Student's paired (difference is observed between paired M_C and M_R values) one-tailed (because $M_C \geqslant M_R$) t-test. Sanderson and Zobel [13] showed that it is more reliable than other tests, such as Wilcoxon's signed rank test. The difference between tested samples is statistically significant when $p < \alpha$, with $\alpha = 0.05$. The smaller p-value, the more significant the difference is [14]. Finally, correlation between samples is reported according to Pearson's r product-moment correlation coefficient for interval scales, and Kendall's τ rank correlation coefficient for ordinal scales.

Effect on Reciprocal Rank. The effect of the chosen reordering strategy on the rank of the first relevant document is shown in Tab. 2. We report reciprocal ranks RR_x truncated to four digits but computations were done using exact values. Rank positions $1/RR_x$ are also presented because they seem helpful for the reader. Table 2 is ordered by descending $\delta_{RC} = 1/RR_R - 1/RR_C$ to focus on most 'lucky' systems. Statistical tests reported in Tab. 5 show a significant difference between RR_C and RR_R. With a Conventional strategy, the first relevant document is significantly ranked higher in the result-list than with a Realistic Strategy although the IRS remains the same. Despite this difference, a strong correlation ($\geqslant 99\%$) exists between the measure values resulting from both strategies except for the *filtering* track, as characterized by a weaker correlation (89%). Overall, RR_C and RR_R values are correlated, showing a slight but significant difference.

Table 2. Top 3 differences between Conventional $1/RR_C$ and Realistic $1/RR_R$ ranks

Track	Result-list	RR_R	RR_C	RR_O	$1/RR_R$	$1/RR_C$	$1/RR_O$	δ_{RC}
ad hoc	padre2 $^{1994}_{195}$	0.0011	0.0667	0.0769	946	15	13	931
	anu5aut1 $^{1996}_{297}$	0.0010	0.0149	1.0000	992	67	1	925
	anu5aut2 $^{1996}_{297}$	0.0010	0.0149	1.0000	992	67	1	925
filtering	antrpohsu00 $^{2000}_{32}$	0.0000	0.5000	1.0000	988	2	1	986
	antrpnohsu00 $^{2000}_{62}$	0.0000	0.0909	1.0000	988	11	1	977
	antrpohsu00 $^{2000}_{62}$	0.0000	0.0909	1.0000	988	11	1	977
routing	cir6rou1 $^{1997}_{118}$	0.0010	0.1429	1.0000	970	7	1	963
	cir6rou1 $^{1997}_{161}$	0.0010	0.0250	0.1429	998	40	7	958
	virtue3 $^{1997}_{228}$	0.0011	0.2000	0.5000	949	5	2	944
web	irtLnut $^{2001}_{516}$	0.0010	1.0000	1.0000	993	1	1	992
	ictweb10nf1 $^{2001}_{525}$	0.0010	0.1667	1.0000	992	6	1	986
	ictweb10nf $^{2001}_{525}$	0.0010	0.1667	1.0000	992	6	1	986

Effect on Average Precision. The three most affected systems regarding AP are shown in Tab. 3, for each track and reordering strategy. Gain between paired strategies is also presented. We focus on gain$_{CR}$, between AP_C and AP_R, which represents the unfair gain obtained by IRSs which benefited from the uncontrolled parameter influencing TREC Conventional reordering. For instance, gain$_{CR}$ reaches 406% for cir6rou1 $^{1997}_{194}$, which deserves $AP_R = 0.0262$ with a fair strategy. It obtained, however, $AP_C = 0.1325$ with the Conventional strategy.

Statistical tests reported in Tab. 5 show a significant difference between AP_C and AP_R for whatever the track. Nevertheless, this difference is small in percentage, which is in line with the observed strong correlation.

Table 3. Top 3 gains between AP_C and AP_R for each 4 tracks

Track	Result-list	AP_R	AP_C	AP_O	gain$_{OR}$ (%)	gain$_{CR}$ (%)	gain$_{CO}$ (%)
ad hoc	ibmgd2 $^{1996}_{291}$	0.0000	0.0001	0.0074	49,867	318	11,858
	issah1 $^{1995}_{246}$	0.0001	0.0003	0.0018	2,718	311	585
	harris1 $^{1997}_{327}$	0.0139	0.0556	0.0556	300	300	0
filtering	IAHKaf12 $^{1998}_{13}$	0.0005	0.0116	0.0116	2,200	2,200	0
	IAHKaf32 $^{1998}_{13}$	0.0005	0.0116	0.0116	2,200	2,200	0
	IAHKaf12 $^{1998}_{39}$	0.0029	0.0625	0.2500	8,400	2,025	300
routing	cir6rou1 $^{1997}_{161}$	0.0000	0.0008	0.0060	11,995	1,435	688
	cir6rou1 $^{1997}_{194}$	0.0262	0.1325	0.2626	902	406	98
	erliR1 $^{1996}_{77}$	0.0311	0.1358	0.5714	1,736	336	321
web	ICTWebTD12A $^{2003}_{15}$	0.0064	0.2541	0.2544	3,861	3,856	0
	irtLnut $^{2001}_{516}$	0.0012	0.0355	0.2667	22,070	2,853	651
	iswt $^{2000}_{490}$	0.0000	0.0004	0.0007	4,248	2,173	91

Table 4. Top 3 gains between MAP_C and MAP_R for each 4 tracks

Track	Result-list	MAP_R	MAP_C	MAP_O	gain_{OR} (%)	gain_{CR} (%)	gain_{CO} (%)
	padre1 [1994]	0.1060	0.1448	0.2967	180	37	105
ad hoc	UB99SW [1999]	0.0454	0.0550	0.0650	43	21	18
	harris1 [1997]	0.0680	0.0821	0.0895	32	21	9
	IAHKaf12 [1998]	0.0045	0.0396	0.0558	1,140	779	41
filtering	IAHKaf32 [1998]	0.0045	0.0396	0.0558	1,140	779	41
	TNOAF103 [1998]	0.0144	0.0371	0.0899	524	158	142
	cir6rou1 [1997]	0.0545	0.0792	0.2306	323	45	191
routing	erliR1 [1996]	0.1060	0.1412	0.2507	137	33	78
	topic1 [1994]	0.2062	0.2243	0.2543	23	9	13
	ictweb10nf [2001]	0.0210	0.0464	0.4726	2,150	121	919
web	ictweb10nf1 [2001]	0.0210	0.0463	0.4660	2,119	120	907
	irtLnut [2001]	0.0102	0.0221	0.2343	2,202	117	960

Effect on Mean Average Precision. The three most affected systems regarding MAP are shown in Tab. 4, for each track and reordering strategy. Values for gain_{CR} are smaller than for AP because several AP values are considered for computing their average (i.e., MAP). Since some result-lists are not skewed by the uncontrolled parameter, the influence of skewed AP values on MAP is limited, as counterbalanced by these non-skewed AP. Despite this smoothing effect due to using the arithmetic mean, we observed unjustified gains yet. For instance, padre1 [1994] earned an extra 37% MAP by only benefiting from the uncontrolled parameter. Thus, without any tangible contribution it was granted $MAP_C = 0.1448$, although it only deserves $MAP_R = 0.1060$ in a unbiased setting. Provided that it had been even luckier, it could have unduly obtained up to $MAP_O = 0.2967$.

Although correlated (Tab. 5), MAP_C and MAP_R values are significantly different. Regarding ranks, however, Kendall's τ shows that IRS ranks computed from MAP do not differ significantly for whatever the track or the reordering strategy. This is due to the fact that difference in MAP is not large enough to change IRS ranks. Moreover, we studied the effect of the tie-breaking strategy (MAP_R vs MAP_C) on the statistical significance of differences between paired systems, for each edition. There are up 9% wrong conclusions: t-test would have concluded to significant differences ($p < 0.05$) with Conventional strategy, but to the contrary with Realistic strategy, and vice versa. As another observation, we found that the rank of 23% of the systems is different when computed on MAP_R or MAP_C. When removing the 25% worst systems, there is still 17% of affected systems. This contradicts the assumption that most ties would have been provided by bad systems. Moreover, we noticed that, for instance, ad hoc uwmt6a0 [1997] was ranked 1st, although containing 57% of ties.

We showed in this section that RR_R, AP_R and MAP_R are statistically different from Conventional counterparts, meaning that there is a noticeable difference between the proposed Realistic fair reordering strategy and TREC's strategy. We discuss the implications of these findings in the next section.

Table 5. Correlation and significance of $M_C - M_R$ ($p < 0.001$ are marked with '*')

Track	RR_C vs RR_R		AP_C vs AP_R		MAP_C vs MAP_R	
	δ_{RC} (%)	corr. r	δ_{RC} (%)	corr. r	δ_{RC} (%)	corr. r
ad hoc	0.60*	0.99	0.37*	1.00	0.37*	1.00
filtering	9.39*	0.89	3.14*	0.99	3.12*	0.99
routing	1.14*	0.99	0.57*	1.00	0.58*	1.00
web	0.55*	1.00	0.40*	1.00	0.45*	1.00

6 Discussion: Tie-Breaking and 'Stuffing' Phenomenon

In Sect. 5.2 we showed that IRS scores are influenced by luck. This is an issue when evaluating several IRSs. Comparing them according to evaluation measures may be unfair, as some may just have been luckier than others. In order to foster fairer evaluations, it may be worth supplying trec_eval with an additional parameter allowing reordering strategy selection: Realistic, Conventional and Optimistic. In the end, other evaluation initiatives based on trec_eval (e.g., NTCIR and CLEF) would also benefit from this contribution.

In addition to the tie-breaking bias, we identified a 'stuffing' phenomenon practiced by several IRSs. At TREC, a result-list is at most comprised of 1,000 documents. We noticed that 7.6% of the studied IRSs retrieve less than 1,000 documents for a topic and 'stuff' their result-lists with documents associated with sim = 0. This is conceptually intriguing: why would a system return an irrelevant document? One rational answer may be: among these stuffed documents some may be relevant and thus contribute to the score, even slightly. And yet, with TREC's current reordering strategy, relevant sim = 0 documents may be top ranked in the 'stuffed' part of the result-list. As a result, they unduly contribute more than if they had been ranked further down the list by the Realistic strategy that we propose. Consequently, it seems mandatory to discourage this 'stuffing trick' aiming to artificially increase measure values. This represents another case for the Realistic reordering strategy that we propose.

7 Related Works

The issue of evaluating runs comprising ties with the common, tie-oblivious, measures (e.g., precision, recall, *F1*, *AP*, *RR*, *NDCG*) was reported in [15,16]. A way to address this issue is the design of tie-aware measures. Raghavan et al. [15] proposed Precall as an extension of precision at varying levels of recall, taking into account groups of tied documents. McSherry and Najork [16] extended the six aforementioned popular measures by averaging over all permutations of tied documents in the result-list. Both of these approaches allow the deterministic comparison of IRS results.

As an alternative solution, we did not design new measures, but tackled the tie-breaking problem by means of reordering strategies applied to the runs instead. The current reordering strategy, that we called Conventional, has been implemented in TREC since its inception. Besides being deterministic, the Realistic and Optimistic strategies that we propose allow the measurement of how much improvement (loss) in effectiveness can be reached when correctly (wrongly) ordering tied documents. A difference between these two bounds can be interpreted as a lack of the system in handling ties properly.

8   Conclusion and Future Work

This paper considered IR evaluation using the trec_eval program, which is used in major evaluation campaigns (e.g., TREC, NTCIR, CLEF) for computing IRS scores (i.e., measure values such as MAP). We underlined that scores depend on two parameters: i) the relevance of retrieved documents, and ii) document names when documents are tied (i.e., retrieved with a same sim value). We argue that the latter represents an uncontrolled parameter influencing computed scores. Indeed, luck may benefit a system when relevant documents are re-ranked higher than non relevant ones, only because of their names.

Counteracting this unfair tie-breaking strategy, we proposed two alternative strategies, namely Realistic and Optimistic reordering. A thorough study of 22 editions of TREC *ad hoc*, *routing*, *filtering*, and *web* tracks showed a statistically significant difference between the Realistic strategy that we propose *vs* TREC's current Conventional strategy for RR, AP, and MAP. However, measure values are not skewed enough to significantly change IRS ranks computed over MAP. This means that the ranking of systems is not affected. We suggest the integration of the two proposed strategies into trec_eval, allowing the experimenter to choose the proper behavior, enabling and fostering fairer evaluations. In addition, this would enable the identification of claimed 'improvements' that only result from chance.

Future work concern three main aspects. First, we plan to test whether the tie-breaking bias affected CLEF and NTCIR, just as it does for TREC. The DIRECT [17] service will be of great help in this respect. Second, as biased evaluation results may have skewed 'learning to rank' approaches [18], it would be worth checking them against fairer evaluation conducted with the proposed Realistic strategy. Third, the study of the 'stuffing' phenomenon discussed in Sect. 6 will quantify the proportion of scores obtained by exploiting side effects related to good knowledge of the evaluation protocol—instead of by improving IRS effectiveness.

Acknowledgments

We thank the reviewers for their thorough evaluations and insightful remarks.

References

1. Robertson, S.: On the history of evaluation in IR. J. Inf. Sci. 34(4), 439–456 (2008)
2. Harman, D.K. (ed.): TREC-1: Proceedings of the First Text REtrieval Conference, Gaithersburg, MD, USA, NIST (February 1993)
3. Voorhees, E.M., Harman, D.K.: TREC: Experiment and Evaluation in Information Retrieval. MIT Press, Cambridge (2005)
4. NIST: README file for trec_eval 8.1, http://trec.nist.gov/trec_eval
5. Kando, N., Kuriyama, K., Nozue, T., Eguchi, K., Kato, H., Hidaka, S.: Overview of IR Tasks at the First NTCIR Workshop. In: Proceedings of the First NTCIR Workshop on Research in Japanese Text Retrieval and Term Recognition, NACSIS, pp. 11–44 (1999)
6. Peters, C., Braschler, M.: European Research Letter – Cross-Language System Evaluation: the CLEF Campaigns. J. Am. Soc. Inf. Sci. Technol. 52(12), 1067–1072 (2001)
7. Clough, P., Sanderson, M.: The CLEF 2003 Cross Language Image Retrieval Track. In: Peters, C., Gonzalo, J., Braschler, M., Kluck, M. (eds.) CLEF 2003. LNCS, vol. 3237, pp. 581–593. Springer, Heidelberg (2004)
8. Voorhees, E.M.: Variations in Relevance Judgments and the Measurement of Retrieval Effectiveness. In: SIGIR 1998: Proceedings of the 21st Annual International ACM SIGIR Conference, pp. 315–323. ACM, New York (1998)
9. Zobel, J.: How Reliable are the Results of large-scale Information Retrieval Experiments? In: SIGIR 1998: Proceedings of the 21st Annual International ACM SIGIR Conference, pp. 307–314. ACM Press, New York (1998)
10. Buckley, C., Voorhees, E.M.: Retrieval System Evaluation. In: [3], ch. 3, pp. 53–75
11. Manning, C.D., Raghavan, P., Schütze, H.: Introduction to Information Retrieval. Cambridge University Press, Cambridge (July 2008)
12. Voorhees, E.M.: Overview of the TREC 2004 Robust Track. In: Voorhees, E.M., Buckland, L.P. (eds.) TREC 2004: Proceedings of the 13th Text REtrieval Conference, Gaithersburg, MD, USA, NIST (2004)
13. Sanderson, M., Zobel, J.: Information Retrieval System Evaluation: Effort, Sensitivity, and Reliability. In: SIGIR 2005: Proceedings of the 28th annual international ACM SIGIR conference, pp. 162–169. ACM, New York (2005)
14. Hull, D.: Using Statistical Testing in the Evaluation of Retrieval Experiments. In: SIGIR 1993: Proceedings of the 16th annual international ACM SIGIR conference, pp. 329–338. ACM Press, New York (1993)
15. Raghavan, V., Bollmann, P., Jung, G.S.: A critical investigation of recall and precision as measures of retrieval system performance. ACM Trans. Inf. Syst. 7(3), 205–229 (1989)
16. McSherry, F., Najork, M.: Computing Information Retrieval Performance Measures Efficiently in the Presence of Tied Scores. In: Macdonald, C., Ounis, I., Plachouras, V., Ruthven, I., White, R.W. (eds.) ECIR 2008. LNCS, vol. 4956, pp. 414–421. Springer, Heidelberg (2008)
17. Di Nunzio, G.M., Ferro, N.: DIRECT: A System for Evaluating Information Access Components of Digital Libraries. In: Rauber, A., Christodoulakis, S., Tjoa, A.M. (eds.) ECDL 2005. LNCS, vol. 3652, pp. 483–484. Springer, Heidelberg (2005)
18. Joachims, T., Li, H., Liu, T.Y., Zhai, C.: Learning to Rank for Information Retrieval (LR4IR 2007). SIGIR Forum 41(2), 58–62 (2007)

Automated Component–Level Evaluation: Present and Future

Allan Hanbury[1] and Henning Müller[2]

[1] Information Retrieval Facility, Vienna, Austria
a.hanbury@ir-facility.org
[2] University of Applied Sciences Western Switzerland (HES–SO), Sierre, Switzerland
henning.mueller@sim.hcuge.ch

Abstract. Automated component–level evaluation of information retrieval (IR) is the main focus of this paper. We present a review of the
current state of web–based and component–level evaluation. Based on
these systems, propositions are made for a comprehensive framework for
web service–based component–level IR system evaluation. The advantages of such an approach are considered, as well as the requirements for
implementing it. Acceptance of such systems by researchers who develop
components and systems is crucial for having an impact and requires
that a clear benefit is demonstrated.

1 Introduction

Information retrieval (IR) has a strong tradition in evaluation, as exemplified by
evaluation campaigns such as TREC (Text REtrieval Conference), CLEF (Cross
Language Evaluation Forum) and NTCIR (NII Test Collection for IR Systems).
The majority of IR evaluation campaigns today are based on the TREC organisation model [1], which is based on the Cranfield paradigm [2]. The TREC model
consists of a yearly cycle in which participating groups are sent data and queries
by the organisers, and subsequently submit retrieval results obtained by their
system for evaluation. The evaluation produces a set of performance measures,
quantifying how each participating group's system performed on the queries with
a stable data set and clear tasks that evaluate the entire system.

This approach has a number of disadvantages [3]. These include:

- fixed timelines and cyclic nature of events;
- evaluation at system–level only;
- difficulty in comparing systems and elucidating reasons for their performance.

These disadvantages are discussed in more detail in Section 2. To overcome
these disadvantages, we suggest moving towards a web–based component–level
evaluation model, which has the potential to be used outside of the cycle of evaluation campaigns. Section 3 discusses some existing approaches to web–based
and component–level evaluation, with examples of systems and evaluation campaigns adopting these approaches. Section 4 presents the promising idea of using

M. Agosti et al. (Eds.): CLEF 2010, LNCS 6360, pp. 124–135, 2010.
© Springer-Verlag Berlin Heidelberg 2010

a web service approach for component–level evaluation. As use by researchers of the existing systems is often lacking, we pay particular attention to motivating participants in Section 5. Long term considerations are discussed in Section 6.

2   Disadvantages of Current Evaluation Approaches

This section expands on the disadvantages listed in the introduction. The first disadvantage is the *cyclic nature of events*, with a fixed deadline for submitting runs and a period of time during which the runs are evaluated before the evaluation results are released. There is usually also a limit on the number of runs that can be submitted per participant, to avoid an excessive workload for the organisers. At the end of each cycle, the data, queries and relevance judgements are usually made available to permit further "offline" evaluation. However, evaluating a system on a large number of test datasets still involves much effort on the part of the experimenter. A solution that has been proposed is online evaluation of systems, as implemented in the EvaluatIR system[1] [4]. This system makes available testsets for download, and allows runs in TREC runfile format (trec_eval) to be uploaded (they are private when uploaded, but can be shared with other users). It maintains a database of past runs submitted to TREC, benchmarks of IR systems and uploaded runs that have been shared, and supports a number of methods for comparing runs. This system not only allows evaluation to be performed when the user requires it, but it makes it possible to keep track of the state–of–the–art results on various datasets.

A further disadvantage is the *evaluation at system–level only*. An IR system contains many components (e.g. stemmer, tokeniser, feature extractor, indexer), but it is difficult to judge the effect of each component on the final result returned for a query. However, extrapolating the effect on a complete IR system from an evaluation of a single component is impossible. As pointed out by Robertson [5], to choose the optimal component for a task in an IR system, alternatives for this component should be evaluated while keeping all other components constant. However, this does not take into account that interactions between components can also affect the retrieval results. For research groups who are experts on one particular component of an IR system, the requirement to evaluate a full system could mean that their component is never fully appreciated, as they do not have the expertise to get a full IR system including their component to perform well.

A further difficulty due to the system–level approach is that when reviewing a number of years of an evaluation task, *it is often difficult to go beyond superficial conclusions based on complete system performance and textual descriptions of the systems*. Little information on where to concentrate effort so as to best improve results can be obtained. Another possible pitfall of the system–level approach, where the result of an evaluation is a ranked list of participants, is the potential to view the evaluation as a competition. This can lead to a focus on tuning existing systems to the evaluation tasks, rather than the scientific goal of determining how and why systems perform as they do. Competitions generally favor small

---

[1] http://www.evaluatir.org/

optimizations of old techniques rather than tests with totally new approaches with a possibly higher potential.

3   Review of Web–Based and Component–Level Evaluation

Several existing campaigns have already worked with component–level evaluation or at least an automated approach to comparing systems based on a service–oriented architecture. The approaches taken are listed below, and each of them is then discussed in more detail:

1. Experimental framework available for download (e.g. MediaMill);
2. Centralised computer for uploading components (e.g. MIREX);
3. Running components locally and uploading intermediate output (e.g. NTCIR ALCIA, Grid@CLEF);
4. Communication via the web and XML based commands (e.g. [6]);
5. Programs to be evaluated are made available by participants as web services (e.g. BioCreative II.5 online evaluation campaign).

For the first approach listed, an experimental framework in the form of a search engine designed in a modular way is made available for download. It can be installed on a local machine, and the modular design should allow simple replacement of various components in the installed system provided that a number of design parameters are satisfied. An example from a related domain is the MediaMill Challenge [7] in the area of video semantic concept detection. A concept detection system, data and ground truth are provided, where the system is broken down into feature extraction, fusion and machine learning components, as shown in Figure 1. Researchers can replace any of these components with their own components to test the effect on the final results. An advantage of this approach is that all processing is done on the local machine. A disadvantage is that the framework always represents a baseline system, so improvements are compared to a baseline, not the state–of–the–art. A solution could be an on-line repository, similar to EvaluatIR mentioned above, which allows results but also components to be shared. Having a system with a fixed number of components and strict workflow also has the problem that new approaches must fit the workflow, limiting the freedom to implement radically new ideas.

The second approach involves making available a server onto which components can be uploaded and incorporated into an evaluation framework running on the server. The components would again have to satisfy various design parameters. A simple way of doing this would be to provide contributors access to the server for uploading and compiling code, which can then be registered in the evaluation framework. Out of necessity due to the copyright restrictions on distributing music, this approach has been adopted by the Music Information Retrieval Evaluation Exchange (MIREX) [8] — one copy of the data is kept on a central server, and participants submit their code through a web interface to be run on the data. The advantage is that the data and code are on the same

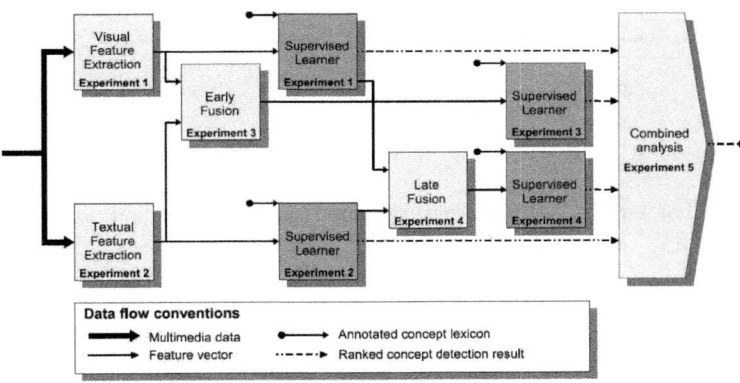

Fig. 1. The concept detection system diagram for the MediaMill Challenge (from [7])

server, so the distribution of huge datasets is avoided, and the evaluation experiments can be run efficiently. The disadvantage for the participants could lie in getting their code to compile and run on a remote server and the risks associated with missing libraries, etc. Participants, in particular companies, could object to uploading proprietary code to a foreign server. Such a system also creates large overheads for the organiser — for MIREX, managing and monitoring the algorithms submitted consumes nearly 1000 person–hours in a year [8].

The Grid@CLEF initiative[2] represents the approach of running components locally and uploading intermediate output. It implemented a component–level evaluation as a pilot track in CLEF 2009 [9]. In order to run these experiments, the *Coordinated Information Retrieval Components Orchestration (CIRCO)* [10] framework was set up. A basic linear framework consisting of tokeniser, stop word remover, stemmer and indexer components was specified (Figure 2). Each component used as input and output XML data in a specified format, the CIRCO Schema. An online system (CIRCO Web) was set up to manage the registration of components, their description and the exchange of XML messages. This design is an intermediate step between traditional evaluation methodologies and a component–based evaluation — participants run their own experiments, but are required to submit intermediate output from each component. The advantages, as pointed out in [9], are that the system meets the component–level evaluation requirements by allowing participants to evaluate components without having to integrate the components into a running IR system or using an API (Application Programming Interface) due to the XML exchange format. This also allows the components to be evaluated asynchonously, although to an extent limited by the necessity of having output from early components available before being able to test later components in the sequence. The disadvantage pointed out in [9] is that the XML files produced could be 50–60 times the size of the original collection, making the task challenging from a computational point of view.

---

[2] `http://ims.dei.unipd.it/websites/gridclef/`

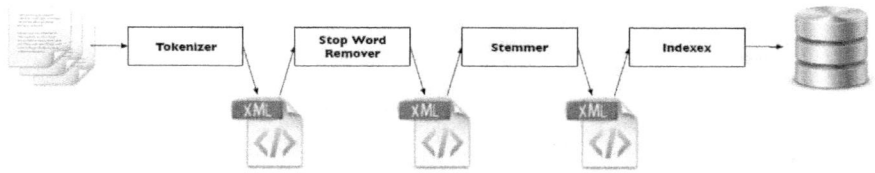

Fig. 2. The Grid@CLEF approach to component–level evaluation (from [9])

A similar approach is adopted in the NTCIR–7 and NTCIR–8 Advanced Cross–lingual Information Access (ALCIA) task[3]. By using the specified XML format, the output from IR modules can be used as input for the Question Answering (QA) modules. It is interesting to note in the NTCIR–7 results [11] that combinations of IR and QA modules from different groups always outperformed combinations from a single group.

In similar ways ImageCLEF has created a nested approach where the output of a first step is distributed to all participants as an input to further steps. In general, textual retrieval results and visual retrieval results are made available to all participants, as not all participants work in both domains. Another step was taken with the Visual Concept Detection Task (VCDT) in 2008, where the results of this task — information on concepts detected in images, such as sky, buildings, animals, etc. — were made available to participants of the photo retrieval task, which used exactly same data set [12]. This gave the participants another source of (potentially noisy) information with which to augment their systems. Unfortunately, few groups integrated this additional information, but those who did showed improved results as an outcome. In the medical task of ImageCLEF 2010 a modality classification task for the entire data set is added before the retrieval phase and results will be made available to participants as for the VCDT task above. Past experiments of the organizers showed that adding modality information to filter the results improved all submitted runs.

An idea for fully automatic evaluation has already been proposed for image retrieval in 2001 [13]. The communication framework (MRML — Multimedia Retrieval Markup Language) was specified, and a web server for running the evaluation by communicating in MRML over a specified port was provided. This system unfortunately did not receive much use as the implementation of a MRML framework would result in additional work for the researchers. The framework also had the disadvantage that the database was fixed (it could of course be extended to several data sets); that due to little use there were few comparisons with state of the art techniques; and that mainly the GNU Image Finding Tool (GIFT[4], [14]) that implements MRML natively was the baseline.

An example of the use of web services in evaluation from a related domain is the BioCreAtIvE (Critical Assessment of Information Extraction systems in Biology) Challenge for annotation in the biomedical domain [15,16]. The approach

---

[3] http://aclia.lti.cs.cmu.edu/ntcir8/
[4] http://www.gnu.org/software/gift/

for BioCreative II.5[5] was to have all participants install a web server to make
their Annotation Servers available. A central MetaServer then calls the available
Annotation Servers that use a standard interface. The Annotation Servers take
as input a full–text document and produce as output the annotation results as
structured data. The advantage of such a system is that researchers can maintain
their tools locally and thus do not need to concern themselves with installation
on another machine. Furthermore, the resources run in a distributed way thus
limiting the charge on a central server. The system also allows the evaluation of
efficiency of the tools, such as the response speed, which is an important criterion
when working with extremely large databases. On the other hand such a system
can favor groups with large hardware budgets.

Table 1 summarises the advantages and disadvantages of the approaches.

4   Towards Web–Based Component–Level Evaluation

To overcome the disadvantages of system–level evaluation, it is necessary to move
IR evaluation towards web–based component–level evaluation. In the BioCreative challenge, complete systems are made available as web services. This is an
approach that would also work in the IR evaluation framework, where participants could expose their search engines through web services. Queries (or finer–
grained tasks) could be sent to these web services by the central Metaserver, and
document lists sent back to the Metaserver for further evaluation. The database
of documents to index could be provided as a download as is usual in evaluation
campaigns, and could later be developed so that documents to index are provided by the Metaserver. This web service–based IR evaluation opens the door
to a component–level evaluation built on the same principles.

A schematic diagram of a general component–based evaluation system is
shown in Figure 3. The basic idea is that a framework consisting of components (an arbitrary framework diagram is shown in Figure 3) is defined, and
contributors can add instances of the components into this framework for use in
running IR experiments. The main challenge is how to instantiate such a framework so that researchers can easily add components to it, and experiments can
be successfully run without creating much additional work for researchers.

The general techniques for developing automatic evaluation systems exist [6]
with the Internet and service–oriented architectures, for example. If all researchers
made their components available via a standardized interface, then these components could all be shared by the various participants and used in many combinations. Already now many IR tools are based on existing components such as
Lucene, Lemur or other open source projects and having service–oriented use of
such tools would simply be one step further. This could even help researchers to
better integrate existing tools and techniques.

Such an evaluation would work as follows. An IR system built out of a set of
components will be specified. Participating groups in the evaluation may choose
which components they wish to submit and which components to use from other

---

[5] http://www.biocreative.org/events/biocreative-ii5/biocreative-ii5/

Table 1. Comparison of the current approaches to component-level evaluation

Approach	Example	Advantages	Disadvantages
Experimental framework download	Mediamill	– Processing done locally	– Framework always represents a baseline system – Fixed workflow
Centralized component upload	MIREX	– Distribution of huge datasets avoided – Evaluation experiments can be run efficiently	– Getting code to compile and run on a remote server – Large overheads for the organisers
Uploading intermediate output	Grid@CLEF, NTCIR ALCIA	– Participants evaluate components without integrating them into complete running systems – Components can be evaluated asynchronously	– Asynchronous evaluation restricted by pipeline structure – XML files produced 50–60 times the size of the document collection
Communication via the web and XML based commands	MRML	– fully automatic and quick evaluation at any time	– Overhead through implementation of the MRML framework – Databases fixed
Web services	BioCreative II.5	– Tools can be maintained locally – Programs run in a distributed way	– Can favour groups with a large hardware budget – Overhead of writing a web service interface to a system

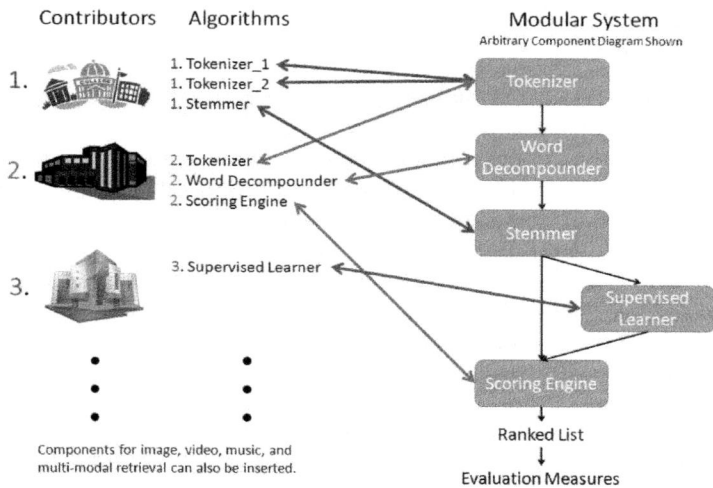

Fig. 3. Overview of a general component–level evaluation framework

groups. These components should be written so as to run on the participants'
computers, callable through a web interface. Participants register their components on a central server. The central server then runs the experiments using a
large number of combinations of components, accessed through their web interfaces. To create such a system, the following are needed:

- software and a central server to run the evaluation;
- protocols for interfacing with programs over the web, exchanging data and
  exchanging results, with the current standard for this being web services, so
  XML–based protocols;
- as for any IR evaluation: large amounts of data, realistic queries and relevance judgements.

The protocol design is the key challenge. The participants' task will shift from
performing the experiments to adapting their code to conform to the protocols.
In order to make this attractive to participants, the protocols should be designed
to have the following properties:

Stability: The protocols should be comprehensively designed to change little
over time — after an initial effort to get their systems compliant, little further
interface work would have to be done by the participants (a standard really
needs to be created).

Simplicity: The initial effort by participants to get their systems compliant
should not be high, as a large initial hurdle could discourage participation.
In addition to a specification, code implementing key interface components
should be provided.

Wide Applicability: Implementing the protocols should enable groups to achieve
more than participation in a single evaluation campaign. Standardizing the

protocols for different evaluation campaigns and potentially for other uses is therefore important.

These properties can be contradictory. A stable protocol that covers all possible eventualities, anticipating all current and future needs, is less simple. A complex protocol could be made simpler by having many optional elements (e.g. the output of a tokeniser could be just a stream of words, or it could include information on word position, sentence boundaries, etc.). With such optional elements, downstream components would require the ability to handle missing elements, or to specify which elements are required so that they can function. Wide applicability can be obtained through the use of a common web service protocol, however many of these protocols do not meet the requirement for simplicity.

For the control software, as the amount of participation increases and the number of components included in the IR system specification increases, the potential number of component combinations will explode. It will therefore not be feasible to test all possible combinations. Algorithms for selecting potentially good component combinations based on previous experimental results and the processing speeds of components, but with low probability of missing good combinations, will have to be designed. It would also be useful for users to have available baseline components with "standard" output to simplify the integration of a new component. Further difficulties to be considered are the remote processing of large amounts of data, where participants with slower Internet connections may be disadvantaged (an initial solution may be to continue distributing the data to be installed locally). It will also be good to ensure that participants with less computing capacity are not at a large disadvantage.

A current problem in IR evaluation that is not addressed at all in this framework is the provision of sufficient data, queries and relevance judgements. With the potential for more efficient experiments, this problem might become worse.

5   Motivating Participation

It is important to design the system so that it is accepted and used by the targeted researchers. The system should be designed so that there are clear benefits to be obtained by using it, even though an initial effort is required to adopt it. The component–level evaluation approach has the following general advantages and benefits:

- A large number of experiments can be executed. Each participant makes available online components, which are then called from a central server. This reduces the amount of work for each participant in running complete IR experiments and allows to reuse components of other participants. More extensive experimental results on component performance can be obtained.
- The best performing combination(s) of components can be identified, where components making up this best performing combination could be from different groups. Different search tasks will also possibly be best performed by different constellations of components allowing even for a query–specific optimization of techniques.

- Significantly less emphasis will be placed on the final ranking of complete systems. The results will be in the form of constellations of which components are best suited for which tasks. It also reduces the perceived competitiveness by removing the ranked list of participants. On the other hand, it is easier for a researcher to have his/her component "win" as there are several categories and not only a best final system.
- Research groups will have the opportunity to concentrate on research and development of those components matching their expertise.
- The reuse of components by other researchers is facilitated. By having other research groups' components available, the building of systems can become easier and other systems using components can increase the number of citations received by publications describing these components.

Despite these general advantages and benefits, there is currently a very low acceptance of component–level evaluation among researchers. For MediaMill, browsing the papers citing [7] gives the idea that while many researchers make use of data and ground truth, few use the system framework. The MRML–based system basically had two users, and there were also only two participants in Grid@CLEF 2009. BioCreative II.5 on the other hand had 15 groups participating showing that such an integration is possible.

When introducing component–level evaluation, the benefits should be made clear through a publicity campaign as a critical mass of participants needs to be reached. It is expected that web service–based systems will become common and thus many researchers might have an interest in such an interface anyway. The main users of such systems will most likely be PhD students and post–docs and for them this can be a much easier start through having a clear framework and not losing time working on already existing parts in poorer quality.

6   Long–Term Considerations and Conclusions

Benchmarking and technology comparisons will remain an important domain to advance science and particularly a domain where good baselines exist and where the application potential is already visible. Such benchmarking has to become more systematic and has to be finer–grained than is currently the case. A compromise also needs to be found leaving researchers the possibility to have totally new approaches but at the same time allowing existing components to be reused. This should make the entire process more efficient and allow researchers to concentrate on the novel parts. Particularly PhD students could benefit extensively from such a concept as the entry burden would be much lower than at present.

Given the additional experimental data that will become available through such a framework, a long–term aim can be to design a search engine that can be built entirely from components based on the task that a user is carrying out and analysis of his/her behaviour (targeted search, browsing, etc.). The ability to clearly see the effect of changes in components on the results of a system should also contribute to solving the problem described in [4]: it is not clear from results in published papers that IR systems have improved over the last decade. The more components that can be called the better the acceptance of such a system will be. A possible implementation of such complex IR systems could be through a workflow paradigm, following the lead of eScience with systems such as Kepler[6] [17]. It might be beneficial to have a centrally managed infrastructure where components can be made available also from groups that lack the computing power to host components. Workflow systems also work in a Grid/Cloud environment [18], which could address the large-scale requirements of a component-based IR system.

There is a large number of challenges that need to be tackled. The problem of obtaining a sufficient number of queries and relevance judgements to allow large scale experiments has to be considered. Innovative approaches to harnessing Internet users for continuously increasing the number of relevance judgements should be examined, such as games with a purpose [19] or remunerated tasks [20]. Furthermore, extremely large databases have now become available but are still only rarely treated by researchers. Another problem is changing databases in which documents are constantly being added and deleted, e.g. FlickR.

A possible first step towards automated component–level evaluation is to create a full system approach for IR evaluation (as in Biocreative II.5). For simplicity, the data should be sent to participants and installed locally as is currently done in evaluation campaigns. Each participant should create a web service interface to their full search system, which can be called by the central server. This will allow research groups to get practice at using the web service approach. Once this approach has been accepted by the research community, the component–level evaluation can be introduced in a stepwise way.

Acknowledgements

This work was partially supported by the European Commission FP7 Network of Excellence PROMISE (258191). We thank the referees who provided a number of excellent suggestions.

References

1. Harman, D.: Overview of the first Text REtrieval Conference (TREC–1). In: Proceedings of the first Text REtrieval Conference (TREC–1), Washington DC, USA, pp. 1–20 (1992)
2. Cleverdon, C.W.: Report on the testing and analysis of an investigation into the comparative efficiency of indexing systems. Technical report, Aslib Cranfield Research Project, Cranfield, USA (1962)
3. Robertson, S.: On the history of evaluation in IR. Journal of Information Science 34, 439–456 (2008)
4. Armstrong, T.G., Moffat, A., Webber, W., Zobel, J.: Improvements that don't add up: ad-hoc retrieval results since 1998. In: CIKM 2009: Proceeding of the 18th ACM Conference on Information and Knowledge Management, pp. 601–610. ACM, New York (2009)

---

[6] http://kepler-project.org/

5. Robertson, S.E.: The methodology of information retrieval experiment. In: Jones, K.S. (ed.) Information Retrieval Experiment, pp. 9–31. Butterworths (1981)
6. Müller, H., Müller, W., Marchand-Maillet, S., Squire, D.M., Pun, T.: A web–based evaluation system for content–based image retrieval. In: Proceedings of the 9th ACM International Conference on Multimedia (ACM MM 2001), Ottawa, Canada, pp. 50–54. ACM, New York (2001)
7. Snoek, C.G.M., Worring, M., van Gemert, J.C., Geusebroek, J.M., Smeulders, A.W.M.: The challenge problem for automated detection of 101 semantic concepts in multimedia. In: Proc. ACM Multimedia, pp. 421–430 (2006)
8. Downie, J.S.: The music information retrieval evaluation exchange (2005-2007): A window into music information retrieval research. Acoustical Science and Technology 29, 247–255 (2008)
9. Ferro, N., Harman, D.: CLEF 2009: Grid@CLEF pilot track overview. In: Working Notes of CLEF 2009 (2009)
10. Ferro, N.: Specification of the circo framework, version 0.10. Technical Report IMS.2009.CIRCO.0.10, Department of Information Engineering, University of Padua, Italy (2009)
11. Mitamura, T., Nyberg, E., Shima, H., Kato, T., Mori, T., Lin, C.Y., Song, R., Lin, C.J., Sakai, T., Ji, D., Kando, N.: Overview of the ntcir-7 aclia tasks: Advanced cross-lingual information access. In: Proceedings of the 7th NTCIR Workshop Meeting on Evaluation of Information Access Technologies (2008)
12. Deselaers, T., Hanbury, A.: The visual concept detection task in ImageCLEF 2008. In: Peters, C., Deselaers, T., Ferro, N., Gonzalo, J., Jones, G.J.F., Kurimo, M., Mandl, T., Peñas, A., Petras, V. (eds.) CLEF 2008. LNCS, vol. 5706, pp. 531–538. Springer, Heidelberg (2009)
13. Müller, H., Müller, W., Marchand-Maillet, S., Pun, T., Squire, D.: A web-based evaluation system for CBIR. In: Proc. ACM Multimedia, pp. 50–54 (2001)
14. Squire, D.M., Müller, W., Müller, H., Pun, T.: Content–based query of image databases: inspirations from text retrieval. In: Ersboll, B.K., Johansen, P. (eds.) Pattern Recognition Letters (Selected Papers from The 11th Scandinavian Conference on Image Analysis SCIA 1999), vol. 21, pp. 1193–1198 (2000)
15. Krallinger, M., Morgan, A., Smith, L., Leitner, F., Tanabe, L., Wilbur, J., Hirschman, L., Valencia, A.: Evaluation of text-mining systems for biology: overview of the second biocreative community challenge. Genome Biology 9 (2008)
16. Morga, A.A., Lu, Z., Wang, X., Cohen, A.M., Fluck, J., Ruch, P., Divoli, A., Fundel, K., Leaman, R., Hakenberg, J., Sun, C., Liu, H., Torres, R., Krauthammer, M., Lau, W.W., Liu, H., Hsu, C., Schuemi, M., Cohen, K.B., Hitschmann, L.: Overview of BioCreative II gene normalization. Gene Biology 9, S2–S3 (2008)
17. Ludäscher, B., Altintas, I., Berkley, C., Higgins, D., Jaeger, E., Jones, M., Lee, E.A., Tao, J., Zhao, Y.: Scientific workflow management and the Kepler system. Concurrency and Computation: Practice and Experience 18, 1039–1065 (2006)
18. Wang, J., Crawl, D., Altintas, I.: Kepler + Hadoop: a general architecture facilitating data-intensive applications in scientific workflow systems. In: WORKS 2009: Proceedings of the 4th Workshop on Workflows in Support of Large-Scale Scienc, pp. 1–8. ACM, New York (2009)
19. von Ahn, L.: Games with a purpose. IEEE Computer Magazine, 96–98 (2006)
20. Alonso, O., Rose, D.E., Stewart, B.: Crowdsourcing for relevance evaluation. SIGIR Forum 42, 9–15 (2008)

The Four Ladies of Experimental Evaluation

Donna Harman[1], Noriko Kando[2], Mounia Lalmas[3], and Carol Peters[4]

[1] National Institute of Standards and Technology (NIST), USA
donna.harman@nist.gov
[2] National Institute of Informatics (NII), Japan
Noriko.Kando@nii.ac.jp
[3] University of Glasgow, UK
mounia@acm.org
[4] Institute of Information Science and Technologies, Italian National Research
Council (ISTI-CNR), Pisa, Italy
carol.peters@isti.cnr.it

The goal of the panel is to present some of the main lessons that we have learned in well over a decade of experimental evaluation and to promote discussion with respect to what the future objectives in this field should be.

1 TREC

TREC was started in 1992 in conjunction with the building of a new 2 GB test collection for the DARPA TIPSTER project. Whereas the main task in the early TRECs was the adhoc retrieval task in English, many other tasks such as question-answering, web retrieval, and retrieval within specific domain have been tried over the years. Online proceedings and other information about TREC can be found at http://trec.nist.gov.

The model for TREC was the traditional Cranfield model for test collections, i.e., a set of documents, a set of questions, and the answers (relevance judgments) for those questions. The 2GB document set was larger than the older test collections by a factor of 1000, i.e., instead of 2 megabytes of abstracts, this collection contained 2 gigabytes of full-text documents. The questions (called topics in TREC) were longer and more structured. The major evaluation issue has been how to obtain the relevance judgments, since it would have been impossible to make the complete relevance judgments that had been done for the earlier collections. Rather than trying a random sample, the decision was made to create the sample by taking the top ranked documents (usually top 100) from the participating groups. This method (known as pooling) worked very well for the first 8 years of the ad hoc track (and most of the other tracks), with several studies made confirming sufficient completeness of the relevance judgments to allow confident re-use of the collection. However as the collections grew larger, such as the web collections, other sampling methods have been needed and the re-usability of these large collections needs further research.

TREC has been running for 19 years now, tackling a wide variety of retrieval tasks such as video retrieval (now TRECvid), CLIR, filtering, and enterprise search. An interactive track ran for 11 years and continued as a subtask in other

M. Agosti et al. (Eds.): CLEF 2010, LNCS 6360, pp. 136–139, 2010.

tracks. TREC 2010 has 7 tracks, including a track for blog retrieval, a legal track, and several tracks working with the 1 billion document ClueWeb09 web data.

2   NTCIR

NTCIR, the Asian version of TREC, started in 1999 and has run in an 18-months cycle. Whereas NTCIR is similar to TREC, there has always been a tighter connection to the NLP community, allowing for some unique tracks. Additionally NTCIR has done extensive pioneering work with patents, including searching, classification, and translation. Online proceedings and more information about NTCIR can be found at http://research.nii.ac.jp/ntcir.

NTCIR-1 worked with 339,483 Japanese and English author abstracts from 65 Japanese academic societies. The 83 Japanese topics were gathered from researchers, with assessments done via pooling and at three levels (relevant, partially relevant and non-relevant). NTCIR-3 (October 2002) used Chinese and Japanese news collection and a smaller Korean one. NTCIR-4 (June 2004) used almost equally sized collections for Japanese, Chinese, Korean and English newspapers published in Asia in the same years to allow complete multilingual testing with four-levels judgments. NTCIR-5 (December 2005) and -6 (May 2007) continued CLIR while adding new documents from 2000-2001.

The NLP connection in NTCIR encouraged both a summarization and a question answering tasks. Single and multiple document summarizations of Japanese news articles were evaluated in NTCIR-2 to -4, trying several evaluation methods. The question answering task started in NTCIR-3 (Japanese only) with a subtask consisting of "followup" questions. This enhanced in NTCIR-4 and -5 as "information access dialog (IAD)" task, simulating an interaction between systems and users. Cross-language QA started in NTCIR-5.

CLIR and Complex CLQA merged into Advanced CLIA task in NTCIR-7 (December 2008) and -8 (June 2010), conducting a component-based evaluation of IR and QA by providing a platform for exchanging interim results from each module and examined the effects of combining modules from different systems.

NTCIR has had a major patent retrieval task since NTCIR-3. The first year the task was technology survey using search topics inspired by newspaper articles in 4 languages against two years of Japanese patent fulltext with the goal of finding relevant patents. It became a true patent invalidity search task in NTCIR-4 to -6 against ten years of Japanese patent fulltext, tested on the sets of 34, 36, 69, 349, 1189, and 1689 topics with different levels of relevance assessment exhaustivity and examined how to get large number of topics by using the rejection citations as the relevant documents. NTCIR-6 focused on CLIR using 10 years of U.S. Patents. This continued to translation task in NTCIR-7 and -8, providing a large Japanese-English sentence alignment corpus derived from 10-years patents. Patent classification have been evaluated, targeting automatic construction of patent maps.

There are other characteristics tasks like WEB in NTCIR-3 to -5 using 1.36 TB collection, opinion analysis, and trend extraction and visualization, etc.

The organization and tasks will be drastically changed in NTCIR-9 for further enhancement of the activities.

3   CLEF

The coordination of the European CLIR task moved from TREC to Europe in 2000 and CLEF (Cross-Language Information Forum) was launched. The objective was to expand the European CLIR effort by including more languages and more tasks, and by encouraging more participation from Europe.

In 2000, CLEF offered the initial four TREC languages (English, French, German and Italian), however, new languages were added each year, up to a total of 13, with representatives from the Germanic, Romance, Slavic and Ugro-Finnic families. In order for a new language to be added, a native-speaking support group needed to obtain adequate newspaper data, have it put into TREC-like formatting, and provide personnel to create topics, perform relevance judgments, etc. As new languages were added, monolingual ad hoc retrieval tasks were proposed. Additionally the cross-language effort (bilingual and multilingual) was continued, with tasks of varying complexity, such as bilingual tasks with English target collections to encourage inexperienced groups using topics in whatever language they prefer, or bilingual tasks with unusual pair of languages where groups could try more advanced techniques, or advanced tasks, such as the multilingual and robust tasks, where groups addressed difficult issues attempting innovative solutions.

There were several issues given the distributed nature of the CLEF evaluation effort. First, the initial topics needed to mirror the types of questions asked within the various countries, and this was done (for example) in CLEF 2001 by asking each of the five language groups that year to generate a set of 15 topics. After a pre-search to check for coverage in each target collection, these 75 initial topics were pruned to a final set of 50 topics which were then translated directly to the other languages, with indirect use of the English master set only when necessary. Note that it is critical that translation takes into account the linguistic and cultural backgrounds of the target languages, i.e., the final topics must represent how a user in a given country would actually ask that question!!

CLEF has attracted many new tracks over the years, such as cross-language question answering (24 groups working with target texts in 9 languages in 2005), and a domain-specific track using structured documents in three languages. One of the most popular tracks is ImageCLEF, which started work with captioned photographs in 2003 and the goal of searching topics built in five languages against the English captions for 30,000 black-and-white photographs. This track expanded to include over 20,000 "touristic" photographs with captions in English, German and Spanish, along with over 200,000 medical images with annotions in several languages.

Working notes from all of the CLEF workshops can be found at http://www.clef-campaign.org/, with formal proceedings produced in the Springer Lecture Notes in Computer Science each year.

4   INEX

The INitiative for the Evaluation of XML retrieval (INEX) started in 2002 to provide evaluation of structured document retrieval, in particular to investigate the retrieval of document components that are XML elements of varying granularity. The initiative used 12,107 full-text scientific articles from 18 IEEE Computer Society publications, with each article containing 1,532 XML nodes on average. The collection grew to 16,819 articles in 2005 and moved on to using Wikipedia articles starting in 2006. Like its sister evaluations, INEX has also had auxiliary "tracks" over the years (see `http://inex.is.informatik.uni-duisburg.de/` for INEX through 2007; the current site is `http://www.inex.otag.nz/`).

The main task of ad hoc retrieval has run since the beginning but with major differences from TREC, NTCIR, and CLEF. Because the goal is the retrieval of a particular element of the document that is demarcated by XML tags, the "relevant" documents can range from a paragraph to the whole document. The general idea has been to present the user with the "best" element in the document with respect to their information request, that is an element that exhaustively discusses the topic without including too many irrelevant topics. The notions of exhaustivity and specificity, although well known to the information retrieval community, are very difficult to measure, and this has caused extensive investigations of new metrics within INEX over the years. This difficulty also extends to the relevance assessments.

The ad hoc topics in INEX have reflected the structural nature of the task. The content-only (CO) topics resemble TREC topics, whereas the content-and-structure (CAS) topics include NEXI query language in the title section, which provides specific structural criteria.

The ad hoc track illustrates the complexity of the evaluation. For instance, in 2007, there were four subtasks with different sets of results: thorough (ranked list of elements), focused (ranked list of focused non-overlapping elements), relevant in context (ranked list of the full documents, but including a set of non-overlapping elements), and best in context (ranked list of the documents, but including the best entry point). Each task models a different user approach, with the focused assuming a top-down look across the ranked elements, and the relevant in context and best in context assuming the user wants to see the full document, but with more complex displays for the elements within that document.

Over the years, INEX has had several additional tracks, with the common thread being the use of XML. The longer running tracks have been interactive (user studies for XML documents) and document mining (clustering and classification of XML structured documents). A book search track for digitized books and entity ranking started in 2007. Additional tracks include Link-the-Wiki track (because the data is Wikipedia), and the efficiency track.

A PROMISE for Experimental Evaluation

Martin Braschler[1], Khalid Choukri[2], Nicola Ferro[3], Allan Hanbury[4],
Jussi Karlgren[5], Henning Müller[6], Vivien Petras[7], Emanuele Pianta[8],
Maarten de Rijke[9], and Giuseppe Santucci[10]

[1] Zurich University of Applied Sciences, Switzerland
martin.braschler@zhaw.ch
[2] Evaluations and Language resources Distribution Agency, France
choukri@elda.org
[3] University of Padua, Italy
ferro@dei.unipd.it
[4] Information Retrieval Facility, Austria
a.hanbury@ir-facility.org
[5] Swedish Institute of Computer Science, Sweden
jussi@sics.se
[6] University of Applied Sciences Western Switzerland, Switzerland
henning.mueller@sim.hcuge.ch
[7] Humboldt-Universität zu Berlin, Germany
vivien.petras@ibi.hu-berlin.de
[8] Centre for the Evaluation of Language Communication Technologies, Italy
pianta@fbk.eu
[9] University of Amsterdam, The Netherlands
derijke@uva.nl
[10] Sapienza University of Rome, Italy
santucci@dis.uniroma1.it

Abstract. *Participative Research labOratory for Multimedia and Multilingual Information Systems Evaluation (PROMISE)* is a Network of Excellence, starting in conjunction with this first independent CLEF 2010 conference, and designed to support and develop the evaluation of multilingual and multimedia information access systems, largely through the activities taking place in *Cross-Language Evaluation Forum (CLEF)* today, and taking it forward in important new ways.

PROMISE is coordinated by the University of Padua, and comprises 10 partners: the Swedish Institute for Computer Science, the University of Amsterdam, Sapienza University of Rome, University of Applied Sciences of Western Switzerland, the Information Retrieval Facility, the Zurich University of Applied Sciences, the Humboldt University of Berlin, the Evaluation and Language Resources Distribution Agency, and the Centre for the Evaluation of Language Communication Technologies.

The single most important step forward for multilingual and multimedia information access which PROMISE will work towards is to provide an *open evaluation infrastructure* in order to support *automation* and *collaboration* in the evaluation process.

M. Agosti et al. (Eds.): CLEF 2010, LNCS 6360, pp. 140–144, 2010.

1   Multilingual and Multimedia Information Access

With a population of over 500 million in its 27 states, EU citizens and companies demand information access systems that allow them to interact with the culturally and politically diverse content that surrounds them in multiple media. Currently, about 10 million Europeans work in other member states of the Union, and alongside the Union's 23 official languages and 3 alphabets, 60 additional regional or group languages are used, as well as approximately one hundred languages brought by immigrants [2]. Most of the Union's inhabitants know more than a single language, and the stated political aim is for every citizen to be able to use their first language and two additional languages in their professional and everyday tasks.

With the advance of broadband access and the evolution of both wired and wireless connectivity, today's users are not only information consumers, but also information *producers*: they create their own content, augment existing material through annotations (e.g. adding tags and comments) and links, and mix and mash up different media and applications within a dynamic and collaborative information space. The expectations and habits of users are constantly changing, together with the ways in which they interact with content and services, often creating new and original ways of exploiting them.

In this evolving scenario, language and media barriers are no longer necessarily seen as insurmountable obstacles: they are constantly being crossed and mixed to provide content that can be accessed on a global scale within a multicultural and multilingual setting. Users need to be able to co-operate and communicate in a way that crosses language and media boundaries and goes beyond separate search in diverse media and languages, but that exploits the interactions between languages and media.

To build the tools of the future that we already see taking shape around us, experimental evaluation has been and will continue to be a key means for supporting and fostering the development of multilingual and multimedia information systems. However this evaluation cannot solely be based on laboratory benchmarking, but needs to involve aspects of usage to *validate* the basis on which systems are being designed, and needs to involve stakeholders beyond the technology producers: media producers, purveyors, and consumers.

Participative Research labOratory for Multimedia and Multilingual Information Systems Evaluation (PROMISE)[1] aims at advancing the experimental evaluation of complex multimedia and multilingual information systems in order to support individuals, commercial entities, and communities who design, develop, employ and improve such complex systems. The overall goal of PROMISE is to deliver a unified and open environment bringing together data, knowledge, tools, methodologies, and development and research communities, as measured by three criteria:

1. Increase in the volume of evaluation data;

---

[1] http://www.promise-noe.eu/

2. Increase in the rate of the utilization of the data by development and research sites;
3. Decrease in the amount of the effort needed for carrying out evaluation and tests.

As a partial consequence of the lowered effort, we expect to find larger uptake and activities in each task. We expect to find a *larger community* of usage, with more sites participating in various scales of experimental participation.

As a third and most important long-term effect we confidently expect to see new and sustainable results emerge from the evaluation process, being deployed fruitfully in research and engineering efforts across the world.

2   Open Evaluation Infrastructure

Evaluation, while often done using high-end computational tools, has mostly been a manual process. This has been a bottleneck for scaling up in volume, but most of all in porting and sharing methodologies and tools. PROMISE aims to help, by providing an infrastructure for the effective automation of evaluation. PROMISE will develop and provide an open evaluation infrastructure for carrying out experimentation: it will support all the various steps involved in an evaluation activity; it will be used by the participants in the evaluation activities and by the targeted researcher and developer communities; it will collect experimental collections, experimental data, and their analyses in order to progressively create a knowledge-base that can be used to compare and assess new systems and techniques. This work will be partially based on previous efforts such as the *Distributed Information Retrieval Evaluation Campaign Tool (DIRECT)* system [1]. The proposed open evaluation infrastructure will make topic creation, creation of pools and relevance assessment more efficient and effective. PROMISE expects noticeable scale increases in each of these aspects. Moreover, it will provide the means to support increased collaboration among all the involved stakeholders, e.g. by allowing them to annotate and enrich the managed contents, as well as to apply information visualization techniques [4] for improving the representation and communication of the experimental results.

3   Use Cases as a Bridge between Benchmarking and Validation

Benchmarking, mostly using test collections such as those provided by CLEF, has been a key instrument in the development of information access systems. Benchmarking provides a systematic view of differences between systems, but only if the basic premises of usage can be held constant. Moving from the task- and topic-based usage for which past information systems have been developed to multilingual and multimedia systems for a diverse user population, these premises are being contested. To better serve the development of future systems, future evaluation efforts need to be explicit about what aspects of usage the systems are expected to provide for, and evaluation needs to be tailored to meet these requirements. To move from abstract benchmarking to more user-sensitive evaluation schemes, PROMISE will, in its requirement analysis phase, formulate a set of *use cases* based on usage scenarios for multimedia and multilingual information access. This will allow future development and research efforts to identify similarities and differences between their project and any predecessors, and address these through choices not only with respect to technology but also with respect to projected usage. This will mean a more elaborate requirements analysis and some care in formulating target notions, evaluation metrics, and comparison statistics.

PROMISE will begin by addressing three initial use cases, and more will be systematically added during the project:

Unlocking culture will deal with information access to cultural heritage material held in large-scale digital libraries, based on the *The European Library (TEL)* Collection: a multilingual collection of about 3 million catalogue records.

Search for innovation will deal with patent search and its requirements, based on MAREC (MAtrixware REsearch Collection): a collection of 19 million patent applications and granted patents in 5 languages.

Visual clinical decision support will deal with visual information connected with text in the radiology domain, based on the ImageCLEF collection with currently 84 000 medical images.

4   Continuous Experimentation

Today, evaluation in organised campaigns is mainly unidirectional and follows annual or less frequent cycles: system owners have to download experimental collections, upload their results, and wait for the performance measurements from the organisers. To overcome this, the open architecture provided by PROMISE will allow for two different modes of operation:

1. The open evaluation infrastructure can be remotely accessed: system owners will be able to operate continuously through standard interfaces, run tests, obtain performance indicators, and compare them to existing knowledge bases and state-of-the-art, without necessarily having to share their own results;
2. In the case of a system implementing a set of standard interfaces, the infrastructure will be able to directly operate the system and run a set of tests to assess its performances, thus speeding up the adoption of standard benchmarking practices.

In this way, PROMISE will automate the evaluation process and transform it from periodic to continuous, radically increasing the number of experiments that can be conducted and making large-scale experimental evaluation part of the daily tools used by researchers and developers for designing and implementing their multilingual and multimedia information systems. The open architecture will also enable commercial developers to use the same evaluation methodologies as research projects do, without necessitating public disclosure of results.

5   PROMISE and CLEF

CLEF is a renowned evaluation framework, which has been running for a decade, with the support of major players in Europe and, in general, in the world. CLEF 2010 represents a renewal of the "classic CLEF" format and an experiment to understand how "next generation" evaluation campaigns might be structured [3]. PROMISE will continue the innovation path initiated for CLEF, will provide infrastructure for CLEF tasks, and will extend the scope of CLEF, through an open evaluation infrastructure; through use-case driven experimentation; through an automated, distributed, and continuous evaluation process and through tools for collaboration, communication, sharing, leading to further community building between the research groups that have turned CLEF into the success that it is.

References

1. Agosti, M., Ferro, N.: Towards an Evaluation Infrastructure for DL Performance Evaluation. In: Tsakonas, G., Papatheodorou, C. (eds.) Evaluation of Digital Libraries: An insight into useful applications and methods, pp. 93–120. Chandos Publishing, Oxford (2009)
2. Commission of the European Communities. Multilingualism: an asset for Europe and a shared commitment. COMM (2008) 566 Final (September 2008)
3. Ferro, N.: CLEF, CLEF 2010, and PROMISEs: Perspectives for the Cross-Language Evaluation Forum. In: Kando, N., Kishida, K. (eds.) Proc. 8th NTCIR Workshop Meeting on Evaluation of Information Access Technologies: Information Retrieval, Question Answering and Cross-Lingual Information Access, pp. 2–12. National Institute of Informatics, Tokyo (2010)
4. Keim, D.A., Mansmann, F., Schneidewind, J., Ziegler, H.: Challenges in Visual Data Analysis. In: Banissi, E. (ed.) Proc. of the 10th International Conference on Information Visualization (IV 2006), pp. 9–16. IEEE Computer Society, Los Alamitos (2006)

Author Index

GPSR Compliance

The European Union's (EU) General Product Safety Regulation (GPSR) is a set of rules that requires consumer products to be safe and our obligations to ensure this.

If you have any concerns about our products, you can contact us on ProductSafety@springernature.com

In case Publisher is established outside the EU, the EU authorized representative is:

Springer Nature Customer Service Center GmbH
Europaplatz 3
69115 Heidelberg, Germany

Batch number: 09478804

Printed by Printforce, the Netherlands